# An Interdependent Approach to Happiness and Well-Being

"This is the book we need right now! Following the first two tumultuous decades of the 21st century, Yukiko Uchida and Jeremy Rappleye invite readers to think broadly and deeply about what is means to be happy. Is the answer wealth, freedom, connection, harmony, health? Is happiness an individual pursuit or is it the larger society that makes happiness possible? How much happiness is the right amount? How can happiness to be measured? How should policymakers in education, sustainability, economic development, and global governance think about the meaning and role of happiness? With a rich and accessible blend of research, data, stories, and descriptions of past and ongoing efforts to capture and chart happiness, the book persuasively demonstrates that happiness depends in very important part on the times and places we find ourselves in. Most of what is currently broadly known about happiness comes from the theories and research by North American researchers grounded in Western philosophical frameworks. Uchida and Rappleye introduce an interdependent model of happiness in which happiness is less about independence, achievement and self-esteem and more about relationships, attunement to others and social support. This approach, prevalent in Japan and much of East Asia, is also relevant to happiness to everywhere, and the authors weave a compelling argument for a focus on interdependence as a potential solution for happiness and for effective solutions to 21st century challenges."

—Hazel Rose Markus, *Stanford University, USA and author of* Clash! How to Thrive in a Multicultural World *(2013)*

Yukiko Uchida • Jeremy Rappleye

# An Interdependent Approach to Happiness and Well-Being

Both authors contributed equally (and
interdependently) to this entire volume

palgrave
macmillan

Yukiko Uchida
Institute for the Future of Human Society
Kyoto University
Kyoto, Japan

Jeremy Rappleye
The University of Hong Kong
Hong Kong SAR

ISBN 978-3-031-26259-3        ISBN 978-3-031-26260-9   (eBook)
https://doi.org/10.1007/978-3-031-26260-9

This Palgrave Macmillan imprint is published by the registered company Springer Nature Switzerland AG. The registered company address is: Gewerbestrasse 11, 6330 Cham, Switzerland

Paper in this product is recyclable.

# Acknowledgments

A volume as ambitious and international in scope as this one inevitably draws from different fields, a range of leading scholars, and works from around the world. So much so that it becomes impossible to acknowledge every contribution within the totality of this wider web. For this we apologize in advance. Still we wish to explicitly acknowledge the pioneering contributions emerging from the field of cultural psychology as a whole, and especially the direction of Shinobu Kitayama and Hazel Markus, Yukiko's inspiring advisors and highly esteemed leaders of the field. Both have spent a lifetime opening the field of psychology in ways that make this volume possible. We also wish to acknowledge the distinguished line of scholars at Kyoto University that have made alternatives viable intellectually, specifically the strong Rinsho and Philosophical Pedagogy traditions in educational studies there. These fields trace roots from the Kokoro Research Center of Kyoto University to Kawaii Hayao, back through Kyoto School philosophers Ueda Shizuteru and Nishida Kitaro. One day this work may be recognized internationally as making pioneering contributions, equally as important to the work of Kitayama and Markus, to opening philosophy and education in similar ways. All this work dovetails with the recent efforts in Anglo-American comparative education circles to open us to alternatives beyond the modern Western paradigm, as currently led by scholars such as Keita Takayama, Hikaru Komatsu, Iveta Silova, and Stephen Carney.

On a more immediate level, Yukiko would like to acknowledge the support of all her former and current colleagues at Kokoro Research Center and the Institute for the Future of Human Society at Kyoto University, and the Center for Advanced Study in the Behavioral Sciences at Stanford University and the Berggruen Institute. She also acknowledge the support of all her former and current lab members, as well as a very long list of collaborators with whom she has published papers leading to this volume, including Kosuke Takemura, Shintaro Fukushima, Hidefumi Hitokoto, Masataka Nakayama, Michael Boiger, Matthias Gobel, Vinai Norasakkunkit, Yuji Ogihara, Igor de Almeida, and Kuba Krys. She also acknowledges friends from her graduate school days who also joined Shinobu Kitayama's lab, including Yuri Miyamoto, Keiko Ishii, Takahiko Masuda, Beth Morling, and Steven Heine. She would also like to acknowledge many other close friends and colleagues from all over the world, including Jeanne Tsai, Batja Mesquita, Ayse Uskul, and Yu Niiya. Each has served as a role model on how to unfold a long, impactful career in the field of cultural psychology. Jeremy is grateful for the opportunity he has received to learn about alternative ways of seeing and being, including the warm welcome to dialogue extended by Takehiko Kariya, Akiyoshi Yonezawa, Kyoko Inagaki, Akiko Hayashi, Nishihira Tadashi, and Satoji Yano, and—now—Yukiko herself. Without their openness to dialogue and years of support, it would have been impossible for a California kid educated at the height of the 'self-esteem' movement to learn to see in new ways. Hazel's generous invitation to dialogue at Stanford (over fabulous dinners and good wine!) for a recent sabbatical was another lesson in attuned interdependence. Acknowledgments must also be made to David Phillips at Oxford who opened a path to the field of comparative education (and comparative social science), within which this whole search finds its meaning. We would also like to recognize that this volume would not have been possible without the path-breaking work of Hikaru Komatsu, who has led the linking of cultural psychology, education, and environmental sustainability. Most of Chap. 6 features work done in collaboration with him.

More personally, Yukiko would like to thank Tadashi for all his support and Sou for all his inspiration. Jeremy now—several decades after the fact—recognizes his father's daily lessons in interdependence, and his Mom for always drawing attention to alternative ways of living and being.

# Contents

# About the Authors

**Yukiko Uchida** is a professor at the Institute for the Future of Human Society, Kyoto University, Japan. From 2019 to 2020, she was a fellow at the Center for Advanced Study in the Behavioral Sciences (CASBS), Stanford University. Upon receiving her Ph.D. in Social Psychology from Kyoto University in 2003, she started her academic career as a visiting researcher at the University of Michigan and Stanford University. As a cultural psychologist, she studies the psychological mechanisms behind the experience of emotions like well-being.

**Jeremy Rappleye** is a professor at the University of Hong Kong, Faculty of Education. He received his PhD in Education from the University of Oxford, and worked for 12 years at Kyoto University, Graduate School of Education in the division of Philosophical Pedagogy. His recent research centers on understanding how diverse institutional patterns (education) derive from different cultural worldviews, with particular focus on conceptualizations of self, (well-)being, and reality.

# Abbreviations

| | |
|---|---|
| EU | European Union |
| GDP | Gross Domestic Product |
| GHC | Global Happiness Council |
| GNI | Gross National Income |
| GNH | Gross National Happiness |
| GSBH | Global Survey of Balance and Harmony |
| HDI | Human Development Index |
| IHS | Interdependent Happiness Scale |
| JCCE | Japan Central Council for Education |
| BLI | Better Lives Index |
| NIYE | National Institute for Youth Education |
| OECD | Organisation for Economic Cooperation and Development |
| PISA | Program for International Student Achievement |
| SPI | Social Progress Index |
| SWLS | Satisfaction with Life Scale |
| UNICEF | United Nations Children's Fund |
| UNESCO | United National Education, Science, and Cultural Organization |
| WHR | World Happiness Report |
| WHO | World Health Organization |

# List of Figures

# 1

# Introduction

What is happiness? What makes people authentically happy? Is it an ecstatic feeling of bliss? Or a calm sense of harmony? Is it fleeting or enduring? Cyclical or progressive? Or is it perhaps less of a moment-to-moment feeling, and more of a cognitive evaluation: reflective retrospection on a job well-done, or even an entire life well-lived? Does happiness derive from a sense of safety and stability? Good health? Being surrounded by a loving family and deep friendships? Or does it arise from looking forward with a sense of hope for the future? Are any one of these factors in isolation sufficient? Or does happiness arise from particular combinations of these diverse elements? And, more broadly, is it the individual that strives to obtain a happy life, or is it an entire society that makes individual happiness possible? While happiness is a common pursuit for the myriad collectives—past and present—that constitute our shared humanity, and among the most enduring research themes across the humanities and social sciences, uncertainties around what it means to 'be happy' abound.

In the modern era, the drive for economic growth has been rooted in our strong belief that material wealth is *the* prerequisite for achieving happiness. Yet, human happiness existed prior to the materialism, infrastructures, and institutions of our contemporary world. For previous generations, or among those living in places not deeply touched by

© The Author(s) 2024
Y. Uchida, J. Rappleye, *An Interdependent Approach to Happiness and Well-Being*,
https://doi.org/10.1007/978-3-031-26260-9_1

modernity, happiness would be found elsewhere. Achieving one's goals in the face of difficult circumstances. Being respected by one's peers or community. Experiencing the joys of family. As such, what most of us now view as indispensable ingredients for a happy life are, upon closer inspection, fairly recent requirements. Happiness must be understood as both something universal and unchanging across the diversity of human experience, and yet happiness is also something dependent on the times and places we find ourselves in. It is both something individuals share in common with others and the preferences that set them apart.

## The Search for Twenty-First-Century Happiness

The first two decades of the twenty-first century have shaken our collective confidence in the patterns of the twentieth century, perhaps most of all the pathways to happiness it presupposed. The twentieth century was defined by rapid economic growth and narratives of progress and development. Growth on all fronts. The twenty-first century is, in stark contrast, being defined by stagnant economic growth, declining birthrates and aging populations, sagging public finances, growing populism and extremism, and a range of global crises, including viral pandemics and— perhaps most 'existential' of all—an accelerating climate crisis. Not expansion, but contraction. Gone is the strong confidence of the twentieth century and the naïve optimism about modern materialism in simple correspondence with happiness.

The symbol of twentieth-century happiness was arguably Gross Domestic Product (GDP), a single index of wealth that—at least in the minds of most policymakers and experts—provided a proxy indicator of happiness ('quality of life') in a given country. The twentieth-century pathway was one paved largely with material wealth, as money could buy the infrastructures, institutions, and entertainments that experts 'knew' would guarantee happiness. Indeed, for most of the twentieth century, attempting to measure happiness in any way different than aggregate GDP was denounced as 'fuzzy' or 'fluffy'. This critique was mainly voiced among economists, those intellectual leaders of the twentieth-century GDP paradigm who had been schooled mainly in the theories of Adam

Smith and Jeremy Bentham, both of whom wrote at the optimistic outset of industrialization in eighteenth-century England.

It is this loss of faith in the twentieth-century patterns that has given rise to a recent explosion of scholarship on happiness, of which this volume itself is a part. Happiness, alongside allied terms such as 'well-being' and 'flourishing', has now emerged to become a major topic of scholarly interest. Debates over happiness and well-being are no longer confined narrowly to philosophy and psychology, but are taking place across the social sciences. Even many economists are now interested, as discussed below. Indeed, a casual online search of the terms 'happiness' and 'well-being' shows a rapid increase in such studies since the late 1990s. This surging interest has washed away the simplicity of the economists' twentieth-century GDP formulation and, at the same time, much of the earlier methodological skepticism that happiness—what initially seems to be merely subjective preferences—could ever become an object of contemplation for 'objective', empirically rigorous social science.

Yet today it is clear that amidst the search to find new models to fit a contracting world, thinkers of all stripes—economists, political scientists, sociologists, linguists, and many others—have shifted to view happiness more objectively. Furthermore, shared interests and collaboration across a range of research fields have given rise to 'happiness researchers', those who have solely focused on the topic and whose work spans disciplinary divides (Diener et al., 2014). In effect, the study of happiness has recently come into its own. Notwithstanding lingering concerns about the validity and reliability of measurement, advances in happiness research and the growing importance of subjective happiness measures over the past decade are undeniable (Oishi, 2009). Even if the twentieth-century GDP=Happiness formula remains strong in the minds of many, the growing addition of auxiliary indicators for happiness and well-being in current research—as reviewed below—arguably represents a crucial turning point for social science. It implicitly demonstrates our collective departure from twentieth-century ways of living, thinking, and being.

## Real World Impact

Nor is this shift narrowly confined to academic research, but has recently exploded in the 'real world' of policymaking as well. This is particularly true over the past ten years, and the trend is evident at virtually all levels—global, national, and sub-national (local). The current situation stands in stark contrast to even the early 2000s, when there were virtually no policy-linked attempts to measure happiness. At the global level, consider just the last decade:

- In 2011, the United Nations adopted resolution 65/309 entitled *Happiness: Towards a Holistic Definition of Development*. This was followed by a UN High Level Meeting entitled *Wellbeing and Happiness: Defining a New Economic Paradigm*. Out of these declarations came a first report outlining the state of happiness worldwide (with an opening chapter on Bhutan's Gross National Happiness), later systematized into the annual World Happiness Reports (WHR). The WHRs annually bring together leading researchers and policymakers around key questions to measure, rank, discuss, and highlight happiness worldwide.
- In 2011, the Organisation for Economic Cooperation and Development (OECD) created its Better Life Index (BLI), which aimed to collect comparative data on aspects of life other than macroeconomic growth measures. This includes indicators of social well-being (e.g., life-satisfaction, education, environmental quality, safety, and civic engagement). The BLI initiative explicitly stated the importance of going 'beyond growth' in the coming decades.
- In 2013, World Health Organization (WHO) issued a report entitled *Promotion of Mental Well-Being: Pursuit of Happiness*, arguing for a shift in focus from mental illness to the promotion of well-being. "The concept of well-being, including mental well-being," it argued, "needs to be operationalized widely a public health strategy" across the world.
- In 2013, academic and business leaders centered around Harvard and MIT, and working closely with Nobel Laureates including Amartya Sen, Douglass North, and Joseph Stiglitz, created the Social Progress Index (SPI). It purposefully excluded economic indices such as GDP,

instead focusing on indicators across three dimensions: 'basic needs', 'foundations of well-being', and 'opportunity'. The European Union (EU) later adopted this Index, producing the first EU Social Progress Index in 2016.

- In 2014, UNESCO Asia-Pacific offices launched a project called 'Happy Schools' with the aim of going beyond academic achievement and reforming schools to focus more on the promotion of happiness and well-being. The initial phase of the project built on Positive Psychology, a US-based movement initiated by the work, *Authentic Happiness* (Seligman, 2004), that aimed to move happiness to the center of educational institutions worldwide. The project has continued, with UNESCO Asia-Pacific recently producing the *Happy Schools Guide and Toolkit* (2021) that is currently being distributed to the 46 member countries of the Asia-Pacific, a region covering more than two-thirds of the world's population.

- In 2015, the OECD utilized its flagship Programme for International Student Achievement (PISA) platform to collect data on the happiness and well-being of 15-year olds worldwide. It inquired about students' life satisfaction, creating a ranking of countries whose school systems produced both high-achievement and high well-being scores. In the PISA 2018 test, the OECD added further questions seeking to understand the frequency of particular emotions in the lives of students, including how often they felt 'happy', 'joyful', 'cheerful', as well as 'sad', 'miserable', and 'scared'.

- In 2017, leaders of the United Arab Emirates and Dubai announced the launch of the Global Happiness Council (GHC), a high-level think-tank aimed at promoting happiness in public policymaking deliberations around the world. Comprised of global leaders in happiness research and development programming, the GHC divided into six sub-councils that advance the happiness in the following policy areas: health, education, personal, workplace, measurement, and city planning. The Council presents its annual reports at the World Government Summit each year, seeking to engage high-level policymakers in a 'Global Dialogue for Happiness'.

- In 2020, UNICEF launched a report entitled *Worlds of Influence: Understanding What Shapes Child Well-Being in Rich Countries*, gathering evidence on 41 member countries of the OECD and/or European Union. It ranked countries according to three categories of 'good childhood'—good mental well-being, good physical health, and skills for life. In 2021, UNICEF followed this with *Understanding Child Subjective Well-Being: a call for more data, research, and policy-making targeting children*. This report announced UNICEF's intention to rollout a larger global survey on child well-being in coming years.

In the wake of these myriad global initiatives, national policymakers have responded. European countries including the UK, France, and Germany have been working on formulating country-level indicators of happiness and well-being. In Japan—a key reference point throughout the current volume—the Prime Minister's Office established an Ad Hoc Research Council on Well-Being in 2011, producing a draft index of well-being indicators. Much of this national measurement work follows the increased demands for global data reporting and comparison, as demanded by organizations such the OECD and UN. That is, the inclusion of well-being indicators in the OECD's PISA studies, for example, has led several countries to discuss the inclusion of well-being measurement and initiatives in schools across the country. For example, in March 2023, Japan's Central Council of Education officially recommended 'well-being' as an education policy focus for the coming decade. Moreover, this momentum at the national level has rippled down to the development of sub-national and even municipal level indicators of well-being. For example, one small city tucked into the mountains in eastern Japan (Hatoyama City), claims to be the 'happiest city' based on these new sub-national comparative indicators. Meanwhile, windswept Aalborg (Denmark) has claimed to be the happiest city in Europe, following results of an EU sponsored happiness survey.

In this way, from a discussion dominated by GDP two decades ago, the world has shifted dramatically toward a discussion centered on happiness and well-being today. Over the past decade, the infrastructures—scholarly, bureaucratic, media, and measurement—have been laid for turning the once 'fuzzy' concept of happiness into a new lodestar for policymaking. Importantly, this shift is not evident merely at the macro-level of overarching policy goals, but is instead coming—as shown in the list above—into more delimited policy domains, including labor policy (workplace well-being), organizational management, urban development, health policy, and—in particular—education. In the current volume, we will focus most of our attention on education, largely for three reasons. First, the domain of education clearly illustrates the strong 'real world impact' that this macro-shift has generated for practice. Second, a focus on education highlights the problems that inhere in this recent shift. Given that education is such a deeply cultural domain, it is an arena that clearly reveals the cultural tensions that are central to this volume. Third, education focuses our attention on the future. That is, when the current shift to happiness is recognized as a departure from twentieth-century models—as we have outlined above—then the creation of a new future beyond GDP=Happiness will inevitably fall on the next generation. What the next generation is being taught today gives us our best glimpse of how happiness will be understood in the future.

If education is one focal point of this volume, the environment is the other. But our analysis seeks to show that the two can be recognized as deeply connected once we adopt an interdependent lens. The accelerating, 'existential' climate crisis unfolding rapidly today is one major cause for the loss of faith in the twentieth-century models. The 2011–2012 high-level UN meetings that generated momentum around the happiness agenda explicitly underscored the unsustainability of the existing 'economic paradigm'. We lead Chap. 6 with this discussion. Several years later, the UN replaced the still optimistic Millennium Development Goals (MGDs) with the more pessimistic Sustainable Development Goals (SDGs), and began talking explicitly about the 'existential' character of the climate crisis. Taking its cue from this, UNESCO has advanced an Education for Sustainable Development (ESD) agenda. Meanwhile the OECD, an organization whose mandate had always been economic growth, has in recent years recognized the climate crisis as the world's foremost challenge. All of this underscores that Happiness≠GDP is being understood within a larger horizon of our highly precarious collective environmental situation. Put simply, the shift away from GDP we outlined above is not simply a pragmatic response to low growth rates, declining birthrates, and weak public finances, but an urgent response to the climate crisis. Even if the eco-existential dimension is not how the general public understands the new happiness and well-being agenda, there is little doubt that intellectual and political elites are turning to face environmental (un)sustainability through this agenda. We wait to address this issue until Chap. 6, in light of the alternative model of happiness and well-being we put forth in the intervening chapters.

## The Problems Inherent in Twenty-First-Century Happiness

Despite our agreement with the general departure from twentieth-century patterns, we still have serious concerns in regards to this recent shift. These concerns motivated us to write this volume. Each new happiness ranking and world report heightens collective interest in happiness and well-being. But—at the same time—it causes the complexities surrounding happiness, those we outlined at the outset, to recede from view. The sequence is by now predictable: national happiness rankings are reported, then headlines and debates ensue for why Country X is so happy, and then everyone rushes to replicate Country X. The headline '*Finland: the happiest Country in the World for Fifth Year Running*' is quickly followed by the next: '*7 Lesson Learned from the Happiest Country in the World*' (these are actual headlines). Rarely asked are the deeper questions: What is being measured? How is happiness being understood? What forms of happiness and well-being are the focus?

Let us again consider the complexities of happiness, but add another layer of complexity: what shapes our perception of happiness? In this volume, our starting point is a realization that while we may experience happiness as an individual feeling, our perception of that experience inevitably reflects the values, norms, history, language, societal zeitgeist, and geographical situatedness of the age and culture we live in. Take, for example, the happiness one derives from spending time in nature versus consumption (e.g., shopping). Those from rural areas derive happiness from these two activities differently from their urban counterparts. The term 'happiness' denotes an ideal state for people living in any situation, but what that 'ideal' signifies and how we go about 'achieving' that ideal differs from culture to culture. This dual recognition of (1) the diversity of views of happiness, and (2) the way culture(s) shapes collective perceptions is what we mean when we refer to a "cultural view of happiness" (Uchida and Ogihara, 2012).

A cultural view of happiness recognizes that the values and views of life that constitute a culture inevitably shape the perceptions of ideal states as well. Depending on the impacts of socio-ecological environment and dominant religious and ethical backgrounds, factors that predate modernity by millennia, the specific forms of happiness that people experience and perceive inevitably differ. Once a given cultural background is firmly in place, subsequent generations consciously or unconsciously reproduce a given cultural view of happiness, learning to answer the ubiquitous question "happiness is _____" (insert here: wealth, freedom, connection, harmony, health) in different ways. A cultural view of happiness recognizes more persistence in perception than twentieth-century Modernization Theory would have us believe: instead of the myriad collectives that constitute humanity sloughing off their cultural skin to emerge as moderns, there is persistence in how different groups see the world and understand modernity. Culture is transmitted through education, language, and institutions, and these cultural institutions and practices persist even today. Recognizing this cultural durability is an important step to understanding how given groups of people perceive happiness differently, their different motivations for happiness, how they seek to obtain happiness in different ways, and even—quite unexpectedly—different views of what an ideal level of happiness may be. Surely everyone wants to maximize happiness, right? Wait until Chap. 4.

Our emphasis on continuity is not intended to deny change. As we will discuss at various parts in this volume, but particularly in Chap. 6, specific cultural definitions of happiness do evolve. For example, one study examined the evolution of the definition of "happiness" in English-language dictionaries (Oishi et al., 2013). In older definitions, the idea that happiness was 'good luck' prevailed. Yet, as modern ways of life gathered pace, English-language definitions of happiness gradually became closer to the idea that happiness is something 'an individual acquires on his or her own'. In contrast, in the case of Japan, and despite all the outward modernization of Japanese society and institutions, even today

Japanese definitions of happiness retain a strong sense of 'being lucky'. In China too, despite an entire twentieth century focused on deep, even radical modernization of the culture (e.g., the Cultural Revolution), the connection between happiness and luck is unshaken, as seen every year in Chinese New Year's Celebrations where across every doorway, even in sparkling modern high-rise apartments, on finds the Chinese characters for 'luck descending' (*fu dao*). Change occurs. Cultures evolve. Yet, any new directions are understood within older frameworks, inevitably shaping those new patterns in ways that represent more cultural continuity than we moderns might expect, and definitely more continuity than the utopian-tinged twentieth-century narratives of progress and development would lead us to believe.

## Our Approach

In the widest sense, the current volume seeks to make visible this cultural view of happiness. As twentieth-century frameworks lose their appeal and the turn toward twenty-first-century happiness gains momentum, our primary aim is to open our readers to some sense of the diversity of patterns of happiness and well-being worldwide. We are concerned that the uniformity in measurement over the past decade is leading us toward uniform policy prescriptions in the coming one. Not only would such policies lack 'fit' in diverse contexts worldwide, but they may even be detrimental for happiness and well-being in some cultural contexts. In Chap. 5, we discuss how, for example, educational policies in Japan aiming for greater 'self-esteem' may actually lower happiness in that context. Moreover, the uniformity in measurement and policy may obscure other forms of happiness useful to address collective global challenges, including the climate crisis. This is an issue to which we dedicate all of Chap. 6.

What we seek to share in this volume is the notion of 'Interdependent Well-Being'. This is a form of happiness and well-being that still remains largely overlooked by most in Western contexts, but is found widely across Japan. We do not intend to claim that this is a specifically Japanese model, only that much of the existing empirical support for the existence of this model comes from Japan. In fact, we believe Interdependent

Well-Being is, in many respects, a dominant foregrounding across much of East Asia. At the same time, it is not a model that is only intelligible to Japanese and other East Asian communities. We are neither indigenous psychologists nor philosophical relativists. Instead, our starting point is a strong belief that Interdependent Well-Being is recognizable in other contexts worldwide, even within Western countries wherein the dominant discourse—particularly in North America—draws attention to a more *Independent* Well-Being mode. Here, at the outset, we also wish to underscore that Interdependent Well-being would only be one among a vast number of other options worldwide. We wholeheartedly endorse recent research that pushes us to look deeper into different forms of interdependence beyond East Asia. Our volume, in addition to whatever it may reveal about Interdependent Happiness and Well-Being, is really an appeal to other researchers worldwide to elaborate a range of additional, alternative models. Only with the diversification of research today will we be able to achieve a level of diversification in thinking, policymaking, and being in the future.

Let us briefly mention something of our paths to collaborate on this volume. Uchida's field of specialization is cultural psychology. Arising out of dialogue between psychology and anthropology, the approach of the burgeoning field of cultural psychology is to empirically examine how a wide range of psychological activities, such as how people think, how they make decisions, how they connect with others, how they perceive themselves, and how they experience emotions, are related to the phenomenon of 'culture'. Research findings across a range of fields have found that how we see things, how we understand human relationships, and how we think about the causes of other people's behavior cannot be understood in isolation from 'culture'. Yet understanding the interaction between the 'macro' social phenomenon of culture and the 'micro' psychological phenomenon of how each individual's mind works is a difficult puzzle, not only for cultural psychologists but for all social science researchers. In the context of this volume, such tensions play out in a central question: How does happiness and emotion relate to the culture in which we live? This question has been Uchida's driving research question for over two decades, a curiosity that initially arose from her interest in richly emotive Japanese classical literature as an undergraduate student at Kyoto University: reading the *Tale of the Genji* and trying to imagine what made Japanese people of eleventh-century Kyoto 'happy'.

Rappleye's fields of specialization spans comparative philosophy, sociology, and education. His approach is to understand the interplay between the pervasive ideas/ideologies of a given society (e.g., philosophy), its institutions (e.g. education systems), and the concrete practices that keep those ideas alive (e.g., school routines and pedagogical models). Numerous studies have shown, for example, just how different philosophy and education are across East Asia, as compared with Western countries. Yet there is little work that links the various fields, despite strong resonance, as we discuss in Chap. 5, between leading themes in Japanese philosophy, dominant institutional patterns in contemporary Japan, and the driving motifs of 'modern' Japanese education. Yet, while Rappleye's approach draws attention to the interplay of various elements of the 'macro' social phenomenon of culture, it has its limitations in understanding how pervasive all this is as the 'micro' level of psychological phenomenon. This is where our transdisciplinary complementarity arises: the 'macro' perspective found in comparative philosophy, sociology, and education complement and are complemented by the 'micro' perspectives offered by cultural psychology. To what extent do the philosophical, social, and educational patterns shape and, in turn, come to be shaped by the 'culture' of a given society? This question has been at the heart of Rappleye's research for much of the past decade, a question arising from the personal disjuncture between being raised in California (an extreme example of independence, even in a Western context) and yet finding 'happiness' in a very different form in East Asia, after living in Japan, Taiwan, and China for two decades (Rappleye, 2020).

Our formal collaboration began around 2017, catalyzed by the release of the OECD's PISA 2015 Student Well-Being Report (OECD, 2017). The PISA 2015 results showed that East Asian countries scored the lowest in 'student well-being' (defined as life-satisfaction) globally. We discuss the details in Chap. 5. Reaching out to Uchida for help in understanding the intricacies of the OECD metrics, Rappleye took his first step into the world of cultural psychology, finding a wealth of research largely missed by educational researchers who conceptualize 'globalization' narrowly in political and sociological terms. Inspired by the richness of cultural psychology, several years later he, through the help of Yukiko, would spend a year-long sabbatical at Stanford University, learning from Hazel Markus and others in the Social Psychology Lab there. Subsequent work has combined the perspectives of cultural psychology and education, as in a 2020 consultancy to 'rewrite' the UNESCO Asia-Pacific Happy Schools project mentioned earlier. Meanwhile, Yukiko was being asked to participate more in policymaking circles in Japan. In 2020 she was invited to become a Member of Japan's Central Council for Education. As the top education policymaking body in the country, she would be asked to translate the findings of cultural psychology into the realm of education policy. Despite different disciplinary foci at the outset, the larger global policy discourses have brought us together. Collaboration has been surprisingly seamless, perhaps a reflection of the fact that the earliest cultural psychology experiments were conducted in Japanese classrooms, catalyzed by Japan's world leading student performance in international comparative tests in the 1980s. At a deeper level, cultural psychology and philosophical pedagogy at Kyoto University trace similar roots: the Kyoto School of philosophy and the pioneering work of those like Kawaii Hayao.

One distinct advantage to the transdisciplinary approach we unfold in this volume is that it allows us to think beyond cultural psychology's central focus on the 'mind'. The standpoint of mainstream psychology has generally been to carefully examine the intra-psychic process of individuals—perception, cognition, emotion, motivation, and action. Cultural psychology pushes mainstream psychology to recognize *inter*-psychic processes, focusing on the ways that contexts, such as culture, influence ways of thinking. Yet, given policy developments over the past decade, it is now imperative that cultural psychology expand further to discuss happiness and well-being in relation to institutional and ideational structures in the wider society, and in relation to new policy movements. What is the 'mindset' *and its concomitant social institutions* that support a happy community? What about a happy workplace? A happy school? What policies and pedagogies would allow for greater happiness and well-being? Understanding better how these more 'macro' structures create the 'micro', and how the 'micro' create the 'macro' is an important new direction for this line of research. Through transdisciplinary collaboration and the urgency generated by 'real world' policy shift toward happiness and well-being, we attempt to make a modest contribution here as well.

## Outline of the Book

Our argument unfolds across the next five chapters. To reiterate, the overarching goal is (1) elaboration of an Interdependent Well-Being approach (or mode) and presentation of (2) evidence to support the assertion that this Interdependent mode is not simply present in Japan/East Asia cultures, but may be an important option for all cultures to engage with, as we collectively face the future. That is, we wish to go beyond mere cataloging of difference in the interest of multiculturalism, and instead move toward a global dialogue and expansive worldwide search for pragmatic policy and practice suggestions. We imagine our primary audience not as specialists in cultural psychology, but instead a wider range of policymakers, experts, social scientists, and media analysts. In short, we want to reach anyone advocating for alternatives to the twentieth-century GDP formulation, but who might not be familiar with decades of work in cultural psychology and the comparative social sciences. With this in mind, we attempt to minimize specialized vocabulary, and provide accessible summaries of recent academic research. As discussed, we return repeatedly to the example of education throughout this volume, as it is an arena that is both widely accessible and highly topical: the meanings, assumptions, and measurement errors we see in education are representative of what is happening across a range of policy and practice domains. In the interest of garnering a more global readership, we have also intentionally minimized references that are too specific (e.g., intra-national variation within Japan). Inevitably some nuance is lost in our decision to privilege accessibility over specificity, and global diversity to local nuance. The same decision was made in relation to the length: we sought to keep this volume compact and concise in the hope that more people will find time to read it.

In Chap. 2, we sketch a world map of happiness and well-being. We begin by addressing several dominant views: that there is a universal model (Maslow's Hierarchy of Needs), that money brings happiness, that happiness is higher in developed countries and so on. We then initiate our discussion of a cultural view of happiness, examining the case of the Satisfaction with Life Scale (SWLS). Our purpose is to provide the historical background, philosophical touchstones, and key definitions that we draw upon throughout the volume.

In Chap. 3, we extend this discussion deeper by digging into definitions of happiness and the ways these definitions are operationalized in measurement. This connects with policymaking, and how policymaking has utilized happiness measures in recent years. Here we return to critically reexamine the policy initiatives around happiness we highlighted previously, showing how nearly all are founded on an *Independent* Well-Being mode. Independent versus Interdependent mode is, again, the key distinction driving the volume as a whole.

In Chap. 4, we lay out the Interdependent Well-Being approach, showings it distinctiveness from an Independent mode. Here we summarize a wealth of findings from cultural psychology, and draw connections to work in philosophy and history that further support the conceptualization. This is the core chapter of the entire volume, elaborating the difference on which the critiques found in Chaps. 2 and 3 make sense, and upon which the implications found in Chaps. 5 and 6 are grounded.

In Chap. 5, we work to link Interdependent Well-Being to the macro cultural contexts. We highlight the Culture Cycle (Markus & Kitayama, 2010), first providing a conceptual model of how cultural institutions and mind co-constitute, and then highlight practices, thus showing how the Interdependent Happiness mode is 'held in place' by the wider cultural ecology. Here we focus specifically on how an Interdependent mode leads to different forms of measurement, different educational practices, and a different view of society and social capital.

In Chap. 6, we turn to explore the future of happiness in the twenty-first century. Here we focus on the pragmatic implications of an Interdependent Well-Being approach, arguing it is worthy of greater discussion because of its links to, among other things, the climate crisis. Herein we share a range of emerging empirical evidence suggesting that shifting toward an Interdependent mode may hasten the return to more sustainable forms of life, and—meanwhile—greater levels of well-being. Much of the research we feature here has just emerged in the past year or two, and gives a preview of how cultural psychology and environmental concerns will overlap in the future.

A brief concluding chapter (Chap. 7) then reviews the argument of the book, points out the limitations of our work, and suggests pathways for future research. Here the key themes become inter-disciplinarity, inter-cultural learning, and urgency.

# References

Diener, E., Oishi, S., & Park, J. (2014). An Incomplete List of Eminent Psychologists in the Modern Era. *Archives of Scientific Psychology, 2,* 20–31.

Markus, H. R., & Kitayama, S. (2010). Cultures and Selves: A Cycle of Mutual Constitution. *Perspectives on Psychological Science, 5,* 420–430.

OECD. (2017). PISA 2015 Results (Volume III): Students' Well-Being, PISA, OECD Publishing, Paris.

Oishi, S. (2009) Shiawase wo kagasuru: shinrigaku kara wakatta koto [*Towards a Science of Happiness: What Psychology Teaches Us*]. Shinyosha.

Oishi, S., Graham, J., Kesebir, S., & Galinha, I. C. (2013). Concepts of Happiness across Time and Cultures. *Personality and Social Psychology Bulletin, 39,* 559–577.

Rappleye, J. (2020). Comparative Education as Cultural Critique. *Comparative Education, 56*(1), 39–56.

Seligman, M. (2004). *Authentic Happiness: Using the New Positive Psychology to Realize Your Potential for Lasting Fulfillment.* Atria Books.

Uchida, Y., & Ogihara, Y. (2012). Bunkateki kofukukan: bunkashinrigakuteki chimi to shourai he no tennbo [A Cultural View of Happiness: Findings and Futures from a Cultural Psychology Approach]. Shinrigaku hyoron [*Japanese psychological review*], 55, 26–42.

# 2

# Happiness: A World Map

Before discussing an alternative approach, we first clarify dominant understandings of happiness and well-being. How is happiness currently being imagined? How does this view of happiness shape views of our own society and the wider world? Why has such a view evolved and become dominant? As discussed in the previous chapter, the past decade has witnessed a remarkable shift away from GDP as proxy for happiness and well-being. Existing assumptions are being critically examined, and new conceptualizations put forth. In this chapter, we aim to sketch the broader historical and intellectual contexts leading up to the current surge in interest around these themes. This discussion sets the stage for the next chapter, wherein we look more deeply at key definitions, new forms of measurement, and—most importantly—inherent limitations therein.

## Are Rich Countries Happier?

The idea that money—material wealth—brings happiness came to dominate our imagination in the twentieth century. Who doesn't want to be rich? Who in wealthy countries hasn't considered themselves lucky to have been born in a 'rich' country? Who would doubt that money gives us the ability to pursue all the things that apparently make us happy:

© The Author(s) 2024
Y. Uchida, J. Rappleye, *An Interdependent Approach to Happiness and Well-Being*,
https://doi.org/10.1007/978-3-031-26260-9_2

better food, a nicer home, a more beautiful partner, a life of ease? The connection between wealth and happiness obviously did not begin in the twentieth century, but the notion that national income could 'index' levels of happiness did. The notion that money equated to happiness, gave rise—when considered in macro terms—to the conceptual formulation that a 'rich country = a happy country'. From this formation was borne the hypothesis that Gross Domestic Product (GDP) could be an index of happiness, a formulation we render shorthand as 'GDP=Happiness'.

While the twentieth century is largely a story of unprecedented economic and material growth, there were wide fluctuations in economic fortunes. These range from the depths of the Great Depression and destruction of the Second World War, to the prosperity of the postwar period. Amid these fluctuations, in 1943 an American psychologist named Abraham Maslow put forth a theory that would come to define mainstream discussions for decades. Maslow's Hierarchy of Needs suggested that people, having once satisfied basic needs (physiological and safety) would gradually turn to seek fulfillment in higher needs (belonging and esteem), as shown in Fig. 2.1. Once the lower needs were met, an individual would then seek for 'self-actualization' at the higher levels.

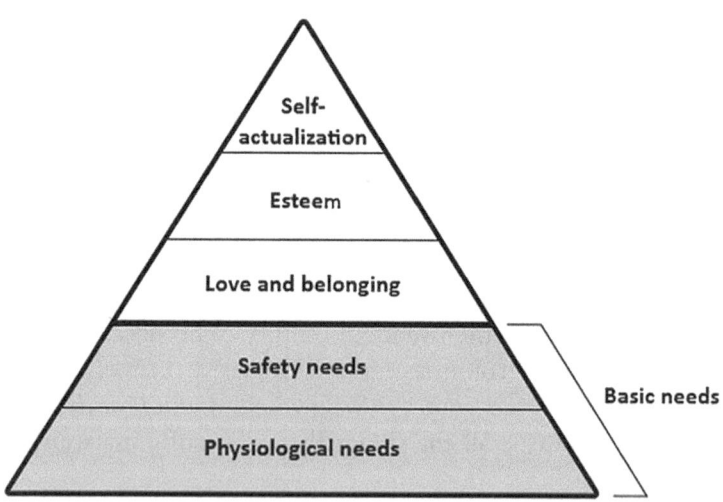

**Fig. 2.1**   Maslow's Hierarchy of Needs

From basic material needs to loftier realms of self-fulfillment. In this way, the material wealth of a society came to be seen as the base upon which all higher needs were built. Material wealth was, in effect, viewed as prerequisite for 'mental wealth'.

Examining historically what occurred in the twentieth century, this formulation makes much sense. In various countries, economic expansion led to a more convenient life. Japan's experience, for example, mirrored that of many countries around the world. After basic needs were met in Japan by the mid-1950s following the devastation of the Second World War, the push was for the so-called 'Three Sacred Treasures' emerged in the late 1950s. Initially, these three treasures were a refrigerator, a washing machine, and a black-and-white television. As the term suggests, affluence brought a sense of happiness. Yet, as the Japanese economy grew further and basic needs were satisfied, more complex needs arose, just as Maslow had predicted. The Three Sacred Treasures were redefined in the 1970s to become an air-conditioner, a color television, and a car, and then redefined again in the 1990s to be a color camera, a DVD player, and a plasma television. From basic safety needs to aesthetic pursuits; from whole family appliances to individual use goods. Indeed, without a basic infrastructure and level of safety guaranteed from the outset, it is difficult for any society to enjoy the 'mental wealth' of happiness. In this sense, it is clear that a certain level of economic wealth—a minimum necessary economic baseline—is important. We would not want to deny this. And, indeed, policymakers thought the same, leading them to pursue this approach via the crude measurement of GDP throughout much of the twentieth century.

Nonetheless, Maslow's Hierarchy of Needs has various problems. These are clear even to non-specialists. Is self-actualization only found among those living in economic affluence? In light of millennia of human beings living together across highly varied economic conditions, is it really accurate to say that the desire to be recognized, respected, and belong arises *only* in high-income contexts? Aren't these drives/needs at play even in difficult economic circumstances? Moreover, is it always that more affluence brings the desire to self-actualize, in the sense of pursuing one's *individual* potential? Isn't it possible that the higher orders lead to further drive for community bonding and belongingness, as in, for example, becoming an integral part of a community? Maslow was American, and it is frequently pointed out that his early formulations of the Hierarchy of Needs tended to privilege the individualism that dominates the North American cultural world (Monnot & Beehr, 2022).

Doubts have also been raised in relation to the extent to which the Maslow model holds. That is, does unlimited economic growth always lead to higher levels of actualization? Is there not some point where economic growth starts to work against happiness? For example, if large numbers of people migrate to urban areas where the infrastructures of convenience allow them to obtain health care and higher paying jobs, but this simultaneously results in higher population densities, cramped living quarters, poor sanitation, air quality, and noise pollution, can we say that their lives are *wholly* better? Is it the case that disconnection from nature, mental stress, and long commute times, resulting in only finding a few hours each week to relax is outweighed by a higher balance in one's bank account? On a more macro, long-term scale, has the rapid expansion of cities and mass consumption, with its tremendous draws on fuel and resources, pushed sustainability out of reach? Even if we ourselves do not pay the price in our lifetimes, the negative effects will be felt by future generations. Maslow did not discuss these sorts of issues. Yet even such casual observations call into doubt his simple formation that material wealth = mental wealth. Even if there was once some element of truth in it, the question moving forward is—as pointed out in the previous chapter—whether that formulation can be sustained in our contemporary world.

These sorts of doubts have come to challenge the twentieth-century World Map of Happiness: the richer the country, the happier we would expect it to be. Slowly but surely, this simple formulation has been discarded. In 1990 the United Nations Development Programme (UNDP) put forth the Human Development Index (HDI). The HDI added additional indicators to GDP (GNI): long and healthy life (measured by life expectancy at birth) and knowledge (mean years of schooling). Moreover, the HDI calculated values and ranked countries around the world, not simply 'developing countries' that was the original remit of the UNDP work. Although 'rich' countries still largely dominated the HDI rankings based on GDP, there was some diversification, as countries with strong health systems and educational attainment moved above countries with higher wealth (e.g., Cuba, as compared with most of Latin America).

Fast forward to today, and we find the OECD Better Life Index (BLI). Launched in 2011, it aims to measure happiness and well-being between countries on 11 indices: housing, income, jobs, community, education, environment, governance, health, life satisfaction, safety, and work-life balance. The OECD introduces the BLI on its website and promotional materials as follows:

> There is more to life than the cold numbers of GDP and economic statistics. This Index allows you to compare well-being across countries based on 11 topics the OECD has identified as essential, in the areas of material living conditions and quality of life. …
>
> …Looking forward, there is no room for complacency. As storm clouds gather on the horizon, mainly from environmental and social challenges, all OECD countries need to take action if they are to maintain today's well-being for future generations. (OECD, 2020, p. 17)

In BLI related materials, the OECD states that there is no country with strengths in all domains. It highlights how countries with high GDP still grapple with problems such as poor work-life balance and depression, for example. What is fascinating here is that the OECD is an organization explicitly formed to advance capitalist development in Europe in the context of the aftermath of the Second World War. It is an institution, born of the geo-political struggles of the twentieth century, dedicated to endless economic growth and unbridled development, in particular economic liberalism. Indeed, these very terms remain enshrined in the very name of the organization. Yet, the OECD itself now realizes that GDP alone cannot index happiness, as evidenced in the creation of the BLI. Moreover, the OECD is aware that growth can be accompanied by a dark side, for example, poor work-life balance and environment degradation. The OECD is thus in the midst of trying to redefine itself as a promoter of a more diversified, less unidimensional approach.

As further indication of the growing doubts around GDP, we may also point to the resounding global success of Bhutan's Gross National Happiness (GNH) initiative. Bhutan is one of the poorest countries in the world. The people of Bhutan are not satisfied with situation and are striving for economic growth. We must be careful not to romanticize the situation there. Nonetheless, there is deeper recognition there that economic wealth does not correspond directly to 'mental' or spiritual wealth. GNH has thus been offered in direct challenge to GDP. Policy decisions are made from a wider perspective, with non-economic conditions accounted for, rather than simply ignored, amid the economic development of the country. There is, to some extent, a recognition that some of what is being lost in economic development are sources of happiness (forests, clean environment, religion, connection). Launched in 2008, Bhutan's GNH policy placed before the world the difference between income and happiness. In fact, the UN's High Level Meeting entitled *Wellbeing and Happiness: Defining a New Economic Paradigm* in 2012, which led to the annual World Happiness Report, was co-chaired by UN Secretary General Ban Ki Moon and Bhutan's then Prime Minister Jigmy Thinley. That such a 'poor' country could be leading a new global discourse underscores just how much a new cartography of global happiness has been taking shape at the outset of the twenty-first century.

## Going Deeper: The Easterlin Paradox

Let us take a deeper look at this. To do so, we turn to research findings that offer greater clarity on happiness worldwide. When we examine the relationship between GDP and country-level subjective happiness indicators, we find that, in fact, higher income countries are more likely to report higher levels of happiness. Figure 2.2 suggests that GDP affects the average level of happiness of a given country, at least to some extent. In the next chapter we will clarify what we mean by 'subjective' indicators and explore different approaches to measurement. For now, our interest remains on the big picture of global happiness and well-being. Similar positive effects between GDP and subjective happiness can be detected at a regional level (Florida et al., 2013; Lawless & Lucas, 2011; Rentfrow et al., 2009). Here then we can confirm that, at least to some extent, economic wealth is important. Note that the data here is from 2008 but the general pattern holds for data from other years as well.

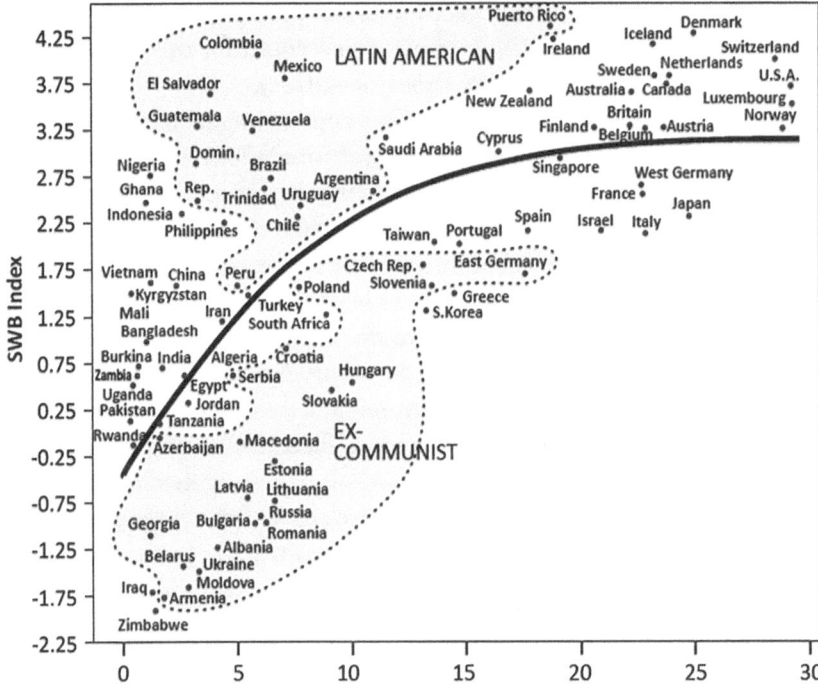

**Fig. 2.2** GDP and subjective well-being (SWB). (Adapted from Inglehart et al., 2008), X-axis is GDP per capita (in $1000 increments)

Interestingly, however, the same relationship does not necessarily hold within a given country. Here then is a paradox, one confirmed through subsequent research studies. In the case of low-income countries, economic indicators and subjective well-being are positively correlated, but when the economy reaches a certain point, the relationship with subjective well-being is no longer linear (Inglehart et al., 2008; Veenhoven, 1999). As stated above, economic indicators and subjective well-being are statistically related to some extent in a country-by-country analysis (i.e., whether the subjective well-being of one country as a whole is higher than that of another country as a whole), but no meaningful relationship may be found in individual differences *within* a country. This has led some researchers to suggest that the effect of economic status on well-being is likely to be due to the "guarantee of individual freedom" that often arises in tandem with greater economic development, and, indeed, when statistical controls are put in place to account for this, the effect of economic status disappears (Fisher & Boer, 2011).

Much of this contemporary research traces its origins back to the initial questions raised within a famous discussion about wealth and well-being unfolding around the 1970s, one that ultimately led to what we now call The Easterlin Paradox. Easterlin, a professor of economics, found that after a certain economic threshold is reached, GDP and subjective happiness no longer correlate. In simple terms: after a certain point, more money does not equate to more happiness. Figure 2.3 shows findings from Japan that support the Easterlin Paradox: from roughly the mid-1970s, levels of happiness have remained constant, despite three decades of economic growth (Cabinet Office, Government of Japan, 2011). Here we see the complexity of simply assuming macro-level indicators (country-level data, i.e., GDP) correlate with micro-level states of mind (individual level data, i.e., subjective well-being). This is a key issue we unpack in the next chapter, when we discuss the methodological challenges of measurement.

How are we to understand this Paradox? Several suggestions have been put forth. First, one explanation underscores that the human mind tends to grow accustomed to its physical and social environment after its

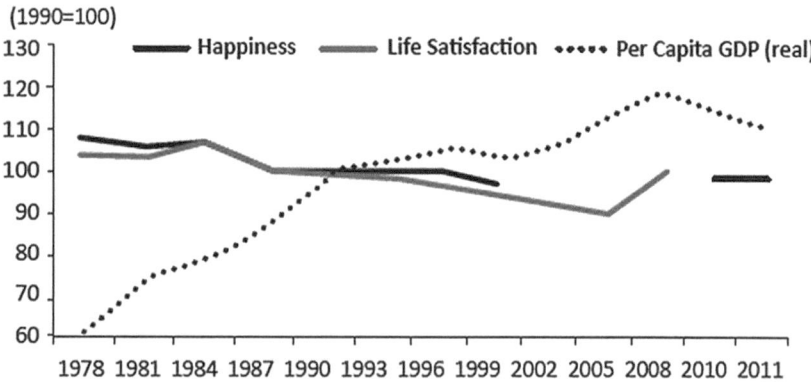

**Fig. 2.3** The Happiness Paradox in Japan (Easterlin, 1974, data adapted from Cabinet Office, Government of Japan, 2011)

'newness' has worn-off. Our sense of happiness is elastic. Viewed negatively, we tend to lose awareness, gratitude, and happiness for novelty after some time. Or, to put it in the language of Maslow outlined above, once our basic needs are met, new desires arise, and the question for belonging, recognition, and self-actualization become the new benchmark for happiness. Viewed in a more positive sense, the human mind adjusts to its surroundings, an insight which may provide important clues for twenty-first-century happiness, as we discuss in Chap. 6.

A second explanation for the paradox is that a rise in GDP tends to create disparities between the so-called winners and losers within a given society. This can lead to a lower sense of trust and fairness among the low-income group(s) and their declining evaluation of their position lowers the sense of well-being as a whole (Oishi et al., 2011). In fact, the growth of GDP in the United States in recent decades is one illustration of this: while GDP figures have increased, only a small number of Americans have shared in that new wealth, and rising inequality has led to a growing sense of dissatisfaction. Similarly, countries such as Argentina and Portugal that have seen growing economic inequality confirm the Easterlin Paradox, whereas in France and Italy which have not experienced similar rises in inequality, evidence for the Paradox is less clear (Oishi & Kesebir, 2015).

A third possibility is that rising GDP is accompanied by negative effects, leading—at some point—to the end of steadily rising happiness. As discussed, the loss of environmental resources comes to mind here. If GDP gains are off-set by pollution, the stresses of urbanization, disconnection (loneliness), and poor work-life balance, then we would expect to find happiness levels 'flatline' at some point, despite further economic growth. Researchers across a range of fields are becoming increasingly aware of this wider context of GDP growth, attempting to understand these connections, and contemplate how to achieve an optimal state of well-being by balancing various factors such as economic conditions, working environment, family environment, and natural environment. The Easterlin Paradox set academic research on a path, one still being unfolded today, of thinking about economic growth as merely one—*but not the only one*—factor influencing relative levels of happiness worldwide.

# Missing Pieces: What Lies Beyond the Subjective?

Scrutinizing Fig. 2.2 a bit more closely, alternative patterns also become evident. While happiness and GDP are positively correlated up to a certain point, there is a relatively narrow dispersion among high-income countries but a high dispersion among low and middle-income countries. Many of the countries whose levels of happiness that exceed what one would normally predict based on GDP figures are located in Latin America. Meanwhile, the countries of Eastern Europe (most former Soviet states) score lower than what one would predict. Among high-income countries, East Asian countries such as Japan score below the average of most of their Western counterparts. Meanwhile, high-income countries in northern Europe such as Denmark score well above average. Indeed, when economic factors are controlled for, happiness is high in countries such as Brazil, Chile, and Argentina, followed by Western countries, and relatively low across East Asian countries such as Japan and South Korea (Diener & Suh, 2000; see also Oishi, 2009). As Inglehart (2008), a prominent happiness researcher who first created Fig. 2.2, points out, these groupings seem to fall along cultural lines, suggesting differences in underlying cultural dispositions.

The Northern European countries have attracted the lion's share of global attention for their high-levels of happiness and well-being. Nordic countries such as Denmark, Iceland, and Sweden consistently top charts like these. The top seven spots in the 2022 World Happiness Report were all taken by Nordic and/or Northern European countries: Finland, Denmark, Switzerland, Iceland, Netherlands, Norway, and Sweden. Finland has topped the WHR for five consecutive years, as touched upon in the headlines we reviewed in the last chapter. In contrast, the low rankings of East Asian countries are equally consistent, with Japan and South Korea scoring among the lowest of countries in the high-income bracket. In light of these rankings, analyses of Nordic countries are abundant, and usually praise is given to the generous social security and welfare provisions (afforded by high taxes), and excellent medical and educational systems. Applying the same analyses in reverse, analyses of Japan and other

East Asia societies come to conclude that the low levels of happiness result from a lack of social security, the weakening of family and community relations, and poor urban planning (high population densities, leading to packed commutes and overcrowded facilities). Levels of happiness in North America—the United States and Canada—are also high, and better than expected based on GDP alone. This general observation has led to, for example, times series analyses, that show that social change (e.g., increased free of choice) is associated with happiness in North America (Inglehart et al., 2008). Building on this, scholars have suggested that countries—not only in North American but everywhere—likely become happier to the extent that they guarantee individual rights and freedoms. The obvious conclusion, albeit often implicit, becomes that countries, such as Japan, do not do well in protecting individuals' rights or extending freedom.

Indeed, research on happiness is an area that lends itself easily to 'comparisons'. Instead of pursuing clear definitions of what happiness and well-being might mean—the contextually based 'contents' of happiness—it is more common to focus on relative comparison, across countries, sub-national units (states, prefectures, cities), occupations, and so on. These are frequent, both in research circles and in media coverage, and tend to produce a policymaking environment that is similarly simplistic in its comparisons. The OECD rankings and the WHR reports are leading examples. These discussions nearly always involve oversimplification of key variables, dubious correlations, and conclusions drawn from comparisons of aggregate values. To be fair, such comparisons may have value in some limited settings and discussions. But more often than not, happiness rankings simplify what is, in essence, a very complex discussion. In the next chapter we turn to critically assess these rankings in detail.

As preview, we may here underscore a few key points. First, virtually all of the measurements that came after GDP have focused on subjective happiness. In other words, measures shifted to asking *individuals* if they feel they are happy. This is true for both Fig. 2.2 and the OECD Happiness Rankings as shown in Fig. 2.4. This move to subjective happiness scales also includes more sophisticated, scholarly attempts, such as the Satisfaction with Life Scale (SWLS) (Diener et al., 1985), which we

| 1. Finland | 15. Belgian | 29. Lithuania |
|---|---|---|
| 2. Norway | 16. Luxembourg | 30. Slovenia |
| 3. Denmark | 17. USA | 31. Latvia |
| 4. Switzerland | 18. Britain | 32. Japan |
| 5. Iceland | 19. Czech Republic | 33. S.Korea |
| 6. Netherlands | 20. Mexico | 34. Russia |
| 7. Canada | 21. France | 35. Estonia |
| 8. New Zealand | 22. Chile | 36. Hungary |
| 9. Sweden | 23. Brazil | 37. Turkey |
| 10. Australia | 24. Spain | 38. Portugal |
| 11. Israel | 25. Colombia | 39. Greece |
| 12. Austria | 26. Slovakia | 40. South Africa |
| 13. Germany | 27. Poland | |
| 14. Ireland | 28. Italy | |

Fig. 2.4   OECD happiness scale ranking. (Adapted from OECD, 2017)

1. In most ways my life is close to my ideal.
2. The conditions of my life are excellent.
3. I am satisfied with my life.
4. If I could live my life over, I would change almost nothing.
5. So far, I have gotten the important things I want in life.

Fig. 2.5   Satisfaction with Life Scale (SWLS) questions. (Adapted from Diener et al., 1985)

examine more closely in the next chapter. For now, the questions that comprise the SWLS are shown in Fig. 2.5. While initially nothing may appear problematic, there is a clear assumption of a particular 'subject' (individual, 'I') these subjective scales attempt to assess. Indeed, all of the scales that are used today were developed in North America, focus on this individual subjective dimension, and produce rankings that find East Asian countries, including Japan, scoring among the lowest. But we must ask—and will, in the pages ahead—can we really just translate measures developed in one culture and use them to access all cultures worldwide?

Implicit in these new happiness scales is also the notion that happiness can be captured in a one-dimensional fashion: that all the world's people respond to a given question in similar ways. It is a well-known phenomenon that in Japan, and other parts of East Asia, there is a cultural response bias to scales like these. Japanese respondents tend to avoid extreme answers at the high and low ends of a scale ('very much' or 'not at all'), in favor of moderate responses ('neither' or 'somewhat'). Meanwhile, differences in reference group also affect responses. For example, when Japanese people access their levels of happiness, they are not imaging comparisons with Americans half a world away, but with other Japanese around them (Heine et al., 2002). This, of course, happens everywhere, not simply in Japan. Subjective valuation requires reference, but these references are not asked about and cannot be held constant. A third problem is that the optimal level of happiness one seeks may differ from country to country. As we shall see in later chapters, the idea that a moderate level of happiness is ideal is dominant in some places of the world, a phenomenon virtually unfathomable for those, like North Americans, that tend to hold the view that the more happiness, the better.

In this chapter, we have attempted to point out that, despite the welcome shift away from GDP toward more nuanced and meaningful indicators over the past several decades, much is still missing. The biggest missing piece in our current attempts at cartographies of happiness and well-being is culture. The general relationship between GDP and happiness appears to be mediated by cultural groupings, response styles affected by culture, and—at a deeper level—optimal levels of happiness differs. All of these pieces are lost when the assumption becomes that all people around the world are, fundamentally, the same in the way they think and feel. Gaps in the existing map are too often overlooked when hasty conclusions are drawn from comparing country averages alone. Most of all, what is missing are detailed analyses of the ways happiness is structured and understood in each country and culture, and what cultural patterns are likely to be associated with higher levels of happiness. If we persist in the current policymaking mood of creating global indicators—a move we do not necessarily oppose—we have to first step back and question what happiness is, rather than quickly transpose existing views of happiness specific to one culture into global measurements, and speculate on how social institutions *everywhere* need to be improved. Yet, as we will now turn to examine, the main global rankings were, indeed, created without accounting for cultural differences in structure and understanding of happiness. That leaves it to us—cultural psychologists and comparative social scientists—to fill those gaps, and locate the missing pieces necessary to redraw the World Map of Happiness.

# References

Diener, E., & Suh, E. M. (2000). *Culture and Subjective Well-Being*. MIT Press.

Diener, E., Emmons, R. A., Larsen, R. J., & Griffin, S. (1985). The Satisfaction with Life Scale. *Journal of Personality Assessment, 49*(1), 71–75.

Easterlin, R. (1974). Does Economic Growth Improve the Human Lot? Some Empirical Evidence. In P. A. David & W. R. Melvin (Eds.), *Nations and Households in Economic Growth* (pp. 89–125). Academic Press.

Fisher, R., & Boer, D. (2011). What Is More Important for National Well-Being: Money or Autonomy? *Journal of Personality and Social Psychology, 101*, 164–184.

Florida, R., Mellander, C., & Rentfrow, P. J. (2013). The Happiness of Cities. *Regional Studies, 47*(4), 613–627.

Heine, S. J., Lehman, D. R., Peng, K., & Greenholtz, J. (2002). What's Wrong with Cross Cultural Comparisons of Subjective Likert Scales: The Reference-Group Problem. *Journal of Personality and Social Psychology, 82*, 903–918.

Inglehart, R., Foa, R., Peterson, C., & Welzel, C. (2008). Development, Freedom, and Rising Happiness. *Perspectives on Psychological Science, 3*, 264–285.

Cabinet Office, Government of Japan. (2011). Measuring National Well-Being—Proposed Well-being Indicators. Available in English online at: https://www5.cao.go.jp/keizai2/koufukudo/pdf/koufukudosian_english.pdf

Lawless, N. M., & Lucas, R. E. (2011). Predictors of Regional Well-Being: A County-Level Analysis. *Social Indicators Research, 101*, 341–357.

Monnot, M. J., & Beehr, T. A. (2022). The Good Life Versus the 'Goods Life': An Investigation of Goal Contents Theory and Employee Subjective Well-Being Across Asian Countries. *Journal of Happiness Studies, 23*(3), 1215–1244.

OECD. (2017). *OECD Better Life Index (2017 Version)*. OECD.

OECD. (2020). *How's Life? 2020—Measuring Well-Being*. OECD. https://www.oecd.org/wise/how-s-life-23089679.htm

Oishi, S. (2009). Shiawase wo kagasuru: shinrigaku kara wakatta koto [*Towards a Science of Happiness: What Psychology Teaches Us*]. Shinyosha.

Oishi, S., & Kesebir, S. (2015). Income Inequality Explains Why Economic Growth Does Not Always Translate to an Increase in Happiness. *Psychological Science, 26*(10), 1630–1638.

Oishi, S., Kesebir, S., & Diener, E. (2011). Income Inequality and Happiness. *Psychological Science, 22*, 1095–1100.

Rentfrow, P. J., Mellander, C., & Florida, R. (2009). Happy States of America: A State-Level Analysis of Psychological, Economic, and Social Well-Being. *Journal of Research in Personality, 43*, 1073–1082.

Veenhoven, R. (1999). Quality-of-Life in Individualistic Society. *Social Indicators Research, 48*, 159–188.

# 3

# Measuring Happiness, Making Policy

In the last chapter, we sketched a general evolution in thinking about happiness, from objective GDP to subjective measurements of individuals. We briefly touched upon some of the potential limitations, but did not go deeply into definitions or forms of measurement. In this chapter, we make that deeper move: first clarifying key definitions that recur throughout this volume, then look at how different forms of measurement feature in contemporary policymaking discussions. The key distinctions are macro/micro and subjective/objective. Within, say, subjective happiness, there are a range of important differences as well, for example, between hedonic and eudaemonic perspectives. Understanding these differences is crucial to lay the groundwork for clarity in thinking about cultural difference in happiness and well-being. This prepares the way for the next chapter, wherein we elaborate the Interdependent Approach.

## Defining Happiness

Without a clear definition of happiness and well-being, there is little hope of measuring and operationalizing them for policymaking. On one level, the definition seems so obvious that it requires little explanation: a positive emotional state that includes a mix of joy and satisfaction. Some

© The Author(s) 2024
Y. Uchida, J. Rappleye, *An Interdependent Approach to Happiness and Well-Being*,
https://doi.org/10.1007/978-3-031-26260-9_3

might even ask: Who needs a definition besides researchers? Isn't 'happiness' that feeling we all recognize when we eat a delicious meal or slip into a warm bath after an exhausting day of work? But researchers might well respond: Do these temporary emotional states fully capture happiness? Doesn't the taste of that meal relate, at least in some ways, to how we look back and evaluate the week, month, or year we have had, in the sense of a luxurious meal as a 'reward' for a month of hard work? For some, doesn't the meal taste all that much better if it is shared with loved ones, spouse, friends, and children, a conviviality unrelated to what is on the menu? And isn't the subjective satisfaction of a hot meal and warm bath made possible by an objective infrastructure of housing and work-life balance? Once again, the complexities of happiness and well-being refuse to go away. So let us formulate some preliminary definitions before moving forward.

## Macro/Micro

In the classical European social science disciplines of sociology, political science, and economics, the focus is often on the macro perspective: what a happy community, society, or polity looks like. Use of economic indicators such as GDP we surveyed in the last chapter is precisely this sort of approach. When European civilization shifted away from a theocentric view during the Enlightenment, it became possible to discuss happiness in 'this life' (as opposed to only the 'next'). Simultaneously, the Enlightenment gave rise to a belief that humans could control these levels of happiness, to some extent, rather than simply waiting on God. This gave rise to new theories such as Bentham's (1789) greatest happiness principle. Bentham defined happiness as the "sum of pleasures and pains" and suggested that morality—and by extension political formulations—should be based on the evaluation of what brings "the greatest happiness for the greatest number" (Veenhoven, 2010). Meanwhile, Adam Smith defined happiness as a layered composite of material wealth, social engagement, and a feeling of tranquility arising from right action (i.e., virtuous conduct) (Matson, 2017). Upon the base of material wealth, individuals had the ability to engage with others, and, in turn, recognize a sense of fulfillment in carrying out one's moral obligations. This progressive structure shares a strong affinity with Maslow's conceptualization discussed in the previous chapter.

In this line of thinking, there has been a strong tendency to conflate macro indicators—that is, the wealth of an entire country or society—with micro indicators—that is, subjective and personal experience. Yet, the 'logic' at play here is not necessarily obvious. For Bentham, a happier society is one in which more happy individuals live. Smith assumed that if a country as a whole is prosperous, then we can conclude that people living in that society must be happy as well. But what about instances when the actions that lead to happiness and pleasure for certain individuals do not necessarily lead to greater happiness for society as a whole? Dumping excess garbage (eliminating excess waste) in public spaces may lead to happiness for certain individuals, but it impoverishes the environment, leading to long-term negative effects for society as a whole. In a different way, there are many cases where individuals sacrifice what they personally want to do for the sake of the greater good. Staying with the example of garbage/waste, if we are somehow obligated to participate in neighborhood clean-up campaigns on, say, a Saturday morning, we may not feel particularly happy about it but it may lead to greater collective well-being.

In this way, the relationship between macro and micro issues is not nearly as straightforward as philosophers and social scientists of the past have tended to portray it. The puzzle of how happiness and well-being arises is, to be sure, as vexing for researchers as it is for the average person. Our point here is merely to provide a vocabulary for the discussion, and also urge caution in conflating the two levels, as we so often see in the classic European social sciences. In this volume, we focus heavily on psychology, which has traditionally been focused on the micro perspective. Yet, we remain very much interested in the macro perspective. What makes a happy community? A happy school? A happy workplace? Much of that discussion comes in Chap. 5. Indeed, in the context of the policy-making shifts we have already described, much recent research focuses on how to think more clearly about the macro dimension, and about the interrelationship between the micro and macro levels.

## Subjective/Objective

With the decline in the belief of GDP as panacea, there has been a con-comitant rise in research on subjective indicators of happiness over the past 20 years, as we discussed. Objective represents an approach marked by measurement or judgment without reference to how those being measured or judged themselves feel, see, or experience the world. Subjective repre-sents the prioritization of precisely the feelings, judgments, and experi-ence of those being measured. The rapid development of social scientific methods and large-scale surveys have driven the recent rise of the subjec-tive perspectives, accumulating an impressive array of empirical findings on happiness and developing increasingly sophisticated survey method-ologies (Oishi, 2009). These have challenged older objective measure-ments, wherein governments or similar official organizations would collect information only on the tangible and/or objectively verifiable. Census data is the classic example here. In sum, the past two decades have finally moved the discussion toward the 'content' of mental wealth.

To think about what the subjective entails, it is important to recognize that subjective judgments are based on an individual's 'cognitive frame'. This cognitive frame has various dimensions and varies depending on situations and context. One of the most common distinctions in contem-porary Western-led research on happiness is between *hedonia* and *eudai-monia*. The former signifies pleasure, enjoyment, comfort, and absence of distress. The latter gestures toward meaning, authenticity, excellence, and a 'living well' sort of evaluation (Huta & Waterman, 2014). The distinc-tion traces its roots back to Aristotle, where in *Nichomachean Ethics* he distinguished *eudaimonia* as a more divine, intellectual, and theoretical sort of happiness, as opposed to the more base, corporeal, and fleeting notion of *hedonia* (see Nagel, 1972). Depending on our cognitive frame, one of these dimensions of happiness comes to be foregrounded.

Building on this, we can also imagine other nuances within cognitive frames. In a study by Oishi (2002), participants in Japan and the United States were asked to rate the intensity of various emotional experiences, including happiness, that they were feeling at a given time of the day over a two-week time span. Participants were subsequently asked to "reflect back over the two weeks" and rate how they felt in general. While there was almost no cultural difference between Japan and the USA in judgments of "current emotions at a given time", there was a marked difference between how Japanese and American respondents felt when they made judgments looking back on the two-week period as a whole. Americans tended to judge the two weeks as having been filled with more good emotions, as compared with Japanese. In other words, there was a culturally induced difference between the emotions one felt at a particular day and the generalized judgments that participants made later. Both feelings at a given instant and judgments over a longer time span—a distinction approximating the difference between *hedonic* and *eudaimonic*—are important for people's decision-making and behavior. In fact, in applied research, as opposed to philosophy, it is more common to distinguish between feelings (daily affect) and judgments of life (life evaluation). The two sets of terms do overlap in some ways. But we need not go into that here. The point is simply that subjective 'cognitive frames' have various dimensions, and also vary by context, producing differences in subjective experience. This latter point is crucial for the overall argument of this volume.

Oftentimes, subjective indicators are deemed less trustworthy than objective ones. This is either because they are seen as simply the 'whim' of the individual and/or not verifiable. Yet, these critiques have been weakened by recognition of larger collective patterns and, at the same time, through the progressive development of increasingly refined indicators and questionnaires. Take, for example, the development of ladder-type scales. The most well-known of these is the Cantril Self-Anchoring Striving Scale (Cantril, 1965). Colloquially, it is simply called the Cantril Ladder. It is frequently used in major international comparative surveys, including the World Happiness Report and Gallup Polls, asking something like the following:

Please imagine a ladder with steps numbered from zero at the bottom to 10 at the top. The top of the ladder represents the best possible life for you and the bottom of the ladder represents the worst possible life for you. On which step of the ladder would you say you personally feel you stand at this time?

What is important about this scale is that it allows respondents to 'self-anchor', meaning they can determine what is important to them, using their own perspective as the foundation of their subjective judgment about their level of happiness. The Cantril Ladder was an attempt to hold the individual's cognitive frame constant, producing greater comparability by having them imagine their own best (many think of wealth, ease, no stress) and worst life (many people think of wartime, famine, disease, uncertainty). Gallup—arguably the world's leading polling organization—also uses this ladder to ask respondents about the future ("On which step do you think you will stand 5 years from now?"). It groups these responses, in combination with other factors, into three groups: Thriving, Struggling, and Suffering (Gallup, 2021). Those in the Thriving category have scores over 7 on the present evaluation, and 8 on future evaluation. Of course, there is much debate about whether this subjective cognitive 'ladder' correctly measures individual happiness. Less frequently, there is debate around whether or not surveys such as these can adequately capture variations in cognitive frames according to context. As we have just seen in the Oishi (2002) study, there is a difference between emotional judgments recorded in real-time and emotional judgments evaluated in retrospect, and focusing on the latter alone would lead to distortions about the differences in levels of happiness between Japan and the USA. We come back around to elaborate on this discussion further in Chap. 5.

The most commonly used subjective measures in psychological research are life satisfaction scales. These attempt to integrate subjective evaluations of life with scales that examine emotional experiences (i.e., asking about the frequency of positive and negative emotional experiences and the intensity of the experiences), and ladder-type self-anchoring scales (Cantril, 1965). There is a vast literature on the reliability of these subjective measures, with much work verifying various levels of validity by showing associations with other behavioral and psychological measures. Among the many, the Satisfaction with Life Scale (SWLS) developed in the 1980s by Professor Ed Diener and his colleagues (Diener et al., 1985), who are now considered to be among the world's foremost authorities on happiness research, is most well-known. We have already seen the items that comprise the SWLS in the previous chapter (Fig. 2.5). The SWLS is known to have good coherence among the five items and high reproducibility when administered over a period of time (Diener et al., 2013; Pavot & Diener, 1993). In addition, there is a certain degree of agreement between self-assessment and peer assessment (Vazire, 2006).

Here, to get a better sense of how all this works, it might be useful for the reader to check their own level of life satisfaction using the scale. Let us reproduce the exact instructions here. Answer each question with a value of 1–7, and then give yourself an overall score by adding together your answers.

---

*Instructions:* Below are five statements that you may agree or disagree with. Using the 1 - 7 scale below, indicate your agreement with each item by placing the appropriate number on the line preceding that item. Please be open and honest in your responding.

- 7 - Strongly agree
- 6 - Agree
- 5 - Slightly agree
- 4 - Neither agree nor disagree
- 3 - Slightly disagree
- 2 - Disagree
- 1 - Strongly disagree

\_\_\_\_ In most ways my life is close to my ideal.

\_\_\_\_ The conditions of my life are excellent.

\_\_\_\_ I am satisfied with my life.

\_\_\_\_ So far I have gotten the important things I want in life.

\_\_\_\_ If I could live my life over, I would change almost nothing.

---

Now let us look at the criteria that Diener et al. attach to the composite score (Diener & Biswas-Diener, 2008), as follows:

31–35—**Extremely Satisfied**: You are satisfied with your life and living circumstances; satisfied with work, family, leisure time, and health.

26–30—**Satisfied**: You are satisfied, but there are elements that could be better; basically a high level of happiness.

21–25—**Slightly Satisfied**: Although the level of satisfaction is generally high, there is room for improvement in some domains; some deviation from the ideal is observed.

20—**Neutral**: Some aspects going well, other not so much.

15–19—**Slightly Dissatisfied**: Unless you answer this questionnaire after some unfortunate event and this situation is simply temporary, there is need for improvement. It may be advisable to change your own ideal state, not your living situation. If you believe that the future will likely be better than the present, there is not a major problem.

10–14—**Dissatisfied**: If you are experiencing temporary depression due to an unfortunate set-back then there is no need for concern, but if that is not the case, you may need to seriously consider ways to improve.

5–9—**Extremely Dissatisfied**: Unless it is a temporary experience of depression, then substantial help and support is needed.

How was your score? Are you 'satisfied' with your life? Are there any of you worried that your score was quite low, perhaps our readers from East Asian backgrounds? Worried that you might be in need of help or changes in the way you live?

Interestingly, one researcher in the UK recently tried to combine the various subjective well-being results from around the world, producing a World Map of Happiness (White, 2007). As shown in Fig. 3.1, virtually

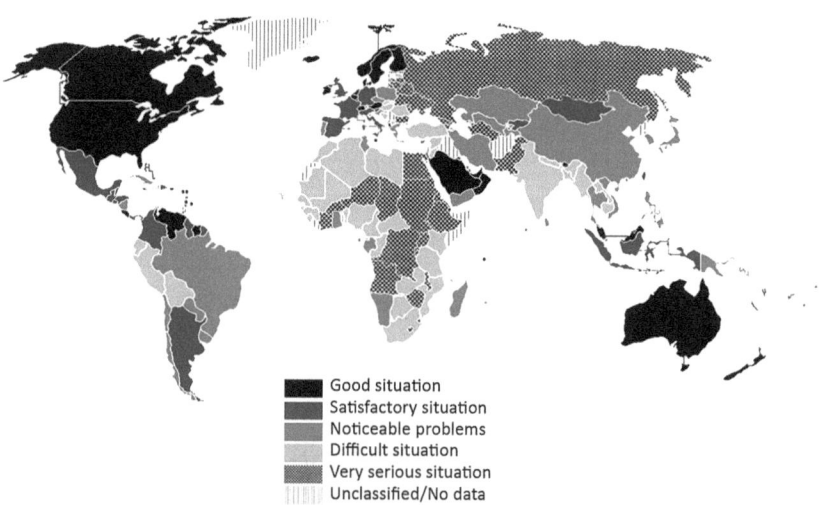

**Fig. 3.1**  World map of happiness. (Adapted from White, 2007)

the whole of East Asia is deemed as having 'noticeable problems', whereas the whole of Anglo-America is in a 'good situation'.

As touched upon in the last chapter, there are cultural differences in happiness. Cognitive frames vary. There are many reasons for this variance, of course. But foremost among these is culture. In fact, scores on the Life Satisfaction Scale tend to be lower in East Asia, especially in Japan. Figure 3.2 derived from Kuppens et al. (2008) shows the five-item averages of scores from 1 to 7 on the vertical axis, as described above. At the top of the scale is Switzerland, where the average score is 5.4, a mark that corresponds to fully satisfied—a high level of overall happiness. The USA does not rank quite as high, but is fairly satisfied. In contrast, the average score for Japan is 3.81, right around the 'slightly dissatisfied' threshold. From this angle, Japan—as a whole—needs improvement, or—perhaps more pessimistically—to lower its expectations.

Yet, here is where things get complicated. Instead of viewing these subjective measures as objective (i.e., understood by everyone everywhere, in the same way), it is important to maintain an awareness of the cognitive frames utilized when creating the measures and when

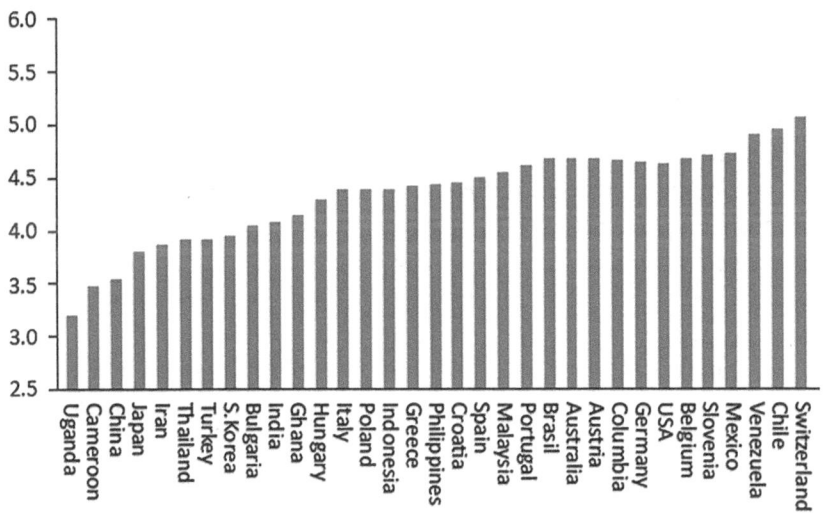

Fig. 3.2 Mean values of the Life Satisfaction Scale. (Based on the values in Table 1 of Kuppens et al., 2008)

responding to those measures. That is, what assumptions implicit in a given cognitive frame lead to the creation of such scales in the first place? How do people with different cognitive frames understand them?

Let us look back at the wording and implicit message of the SWLS. Here we see that the questions convey the sense of personal acquisition: "So far I have gotten the important things I want in life". This suggests the notion of achievement, moving toward goals and obtaining something. Perhaps elements of this happiness formula are found in all cognitive frames, but the dominance of these elements—that is, the degree to which this possessive element is foregrounded—is by no means universal. In the Japanese case, notions of happiness tend to foreground a sense of tranquility and conviviality with significant others.

The SWLS scale is dominated by ratings based on personal acquisition, such as "I have gotten what I wanted so far". Of course, personal achievement is part of the Japanese definition of happiness as well. On the other hand, however, perhaps the SWLS does not capture the Japanese sense of well-being very well. In Japan, happiness tends to be understood as something peaceful, non-active (quiescence), ordinary (as opposed to extraordinary), and/or realized not only by oneself but also with others. *It is less acquisition than attunement.* We shall explore all this further in the next chapter, but the point here is to become aware of the differences, both in assumptions and outcomes, still at play even within the subjective happiness shift.

Of course, we do not wish to give the impression that the entire discussion around happiness has shifted to subjective indicators after the departure from the objective GDP indicator. Paralleling new research on the subjective side, there have been many novel attempts to create better objective indicators. At the global level, the OECD's Better Lives Initiative (BLI) consists primarily of a range of objective indicators, including: income (measured by net financial wealth), jobs (long-term unemployment rate), housing (rooms per person), work-life balance (time devoted to leisure and personal care), health (life expectancy at birth), education (educational attainment; students' reading skills), environmental quality (air pollution), safety (homicide rate; assault rate) and so on (OECD, 2018). Below we delve deeper into the OECD's BLI, but here we can see how an aggregate of objective indicators has replaced GDP in some of the new exercises. Researchers working globally have also examined objective measures, including tax status (Oishi et al., 2011), local social capital (Helliwell & Putnam, 2004; Ram, 2010), and levels of inequality (several). Similar trends toward diversification of objective measures can be found within specific countries. In Japan, for example, one prominent survey carried out by researchers at Hosei University extracted figures from existing socioeconomic statistical databases posited as indicative of elements of happiness for respondents, including: birth rates, homeownership rates, nursery school capacity, job turnover, working hours, number of traffic accidents, and so on. Much of the development of these sorts of indicators tend to be prominent in fields such as public policy, economics, and sociology—all fields that are primarily focused on improving social conditions and are less familiar with psychological surveys.

Undoubtedly, this 'objective' work is important and welcome. Yet, it is crucial to also keep questioning the validity of the implicit judgment that a particular objective indicator—say, ownership of a car or home—actually brings happiness. The 'meaning' of a car and its relation to happiness obviously depends on local contexts, specific historical period, and a range of other factors. Having a car in a congested urban space with highly developed public transportation networks would obviously mean something very different than having a car in the deep countryside where the bus comes twice a day and the nearest supermarket is ten kilometers away. Similarly, being the first in one's family or community to own a car would mean something vastly different than living in a time when virtually everyone has a personal automobile. An over-reliance on objective indicators implicitly imposes a uniform definition on the complexity of happiness and well-being. On the other hand, the sole use of subjective indicators tends to overlook larger trends that can be addressed through public policy. If, for example, subjective life-satisfaction surveys show high levels of happiness, but nonetheless levels of inequality (e.g., the United States) or suicide rates in young adults (e.g., Finland) are high, then we can hardly claim that all is well. For this reason, there is an increasing trend within policymaking circles to combine subjective and (diversified) objective indicators, creating composite happiness indices. In a moment we will look at some of these. Simultaneously, among researchers, there has been a greater push—coinciding with the past few decades of greater attention to subjective indicators—to improve statistical analyses that combine the two. Developments in multilevel modeling and analyses have made it increasingly possible to understand the relationship between objective and subjective phenomena, and the relationship between macro indicators such as economic status (GDP) and economic disparity (Gini coefficient) and micro indicators such as an individual's sense of well-being (Oishi et al., 2012).

## Measuring Happiness, Making Policy

Having clarified the key terms and tensions, let us now circle back around to contemporary policy trends aimed at measuring happiness and well-being. As we saw previously, the past decade alone has witnessed the creation of the WHR, the OECD's BLI, and a plethora of derivative surveys, particularly in the field of education, such as the 2020 UNICEF Report *Child Well-being in Rich Countries*. These global movements have been mirrored by the creation of national level indicators across many countries. Countries as diverse as UK, Israel, and Thailand are now measuring happiness. This shift has occurred, in large part, because the longstanding skepticism by hard-nose policy analysts over the 'fuzzy', subjective nature of happiness has been broken through by marked achievements by social scientists, particularly psychologists. That is, the methodological rigor, analytical sophistication, and empirical accumulation of psychologists have opened space for serious discussions around these themes and inclusion within policymaking deliberations, even among the traditionally skeptical economists (e.g., Frey, 2018).

At a deeper level, there have always been concerns about the possible misuse of happiness as an official indicator. The arguments goes like this. First, if happiness is subjective and unique to an individual, then it is not the role of the state to intervene. Second, measurement may produce clarity, but it can also lead to control. On both of these points we sympathize greatly. But we also come out in favor of the shift to include happiness in policymaking. In terms of the first objection, happiness is indeed subjective to some extent and many of the links between the subjective and objective remain opaque, as we discussed above. It follows then that, without understanding the precise linkages, policy interventions would be imprecise at best, distorting at worst. Moreover, the fact is that states already collect data on a range of behavioral indicators and behind this is the assumption that particular subjective states exist. Take, for example, suicide, a topic for which official statistics already existed in the late nineteenth century (making possible Durkheim's *Suicide: A Study in Sociology* (1897)). Suicide is seen as objective, and there is little skepticism around whether these statistics capture the full complexity of the problem. What drives the collection of suicide figures is the implicit assumption that these indicate an overall social phenomenon of dis/satisfaction with a given society. There is an assumption this behavior corresponds to decisions made from a particular subjective state of mind. The only real difference between a behavioral indicator like suicide and a subjective indicator like happiness is how difficult it is to observe from 'outside'. But, in fact, numerous studies over the past two decades have shown how subjective variables can predict subsequent behavior.

In terms of the second objection—that measurement leads to control—we certainly recognize the historical precedents and are cognizant of contemporary policy trends. The past several decades have seen the rise of neo-liberal models of governance and policymaking. Those skeptical of the shift to happiness do not see these two as unrelated: a shift to subjective happiness distracts attention away from rising inequality, lower growth rates, and the breakdown of public services—the argument goes. Writing in the field of education, Vintimilla (2014) argues that neo-liberalism inevitably favors a pedagogy of fun and blissful happiness, thus pushing out other objectives, including social justice themes. Adams et al. (2019) argue that the mainstream shift to happiness arises from and contributes to furthering neo-liberalism because it treats individuals as separate from context—economic, cultural, ecological, and so on. In other words, mainstream policymaking and even much academic work look at individuals in the same way that neo-liberal policymakers do: as capable of generating happiness by changing their mindset alone. Here Positive Psychology, and its educational spin-off, Positive Education, may immediately come to the reader's mind: these movements tend to focus on mindsets, and rarely challenge the frameworks of individualism that underpin policy neo-liberalism.

Viewed in this way, the shift to 'happiness' is not a new frontier for policymaking, but therapy for a world broken by neo-liberalism. Psychology is simply pushing us inward, encouraging us to 'be happy' as a form of therapy (Madsen, 2014), thus blunting a critical response. While we understand and sympathize with these critiques, we do not share them. As will become clear in subsequent chapters, we are working against the notion that individuals can change their mindset, and also work against an a-contextual view of happiness and well-being. In fact, the alternative approach we are attempting to make visible in this volumes reminds us of context, leads to institutions that are conducive to increasing social connections, and recognizes the ill-effects of hypercapitalism and neo-liberal subjectivity, on societies and the environment. In this alternative model, the happiness of the 'individual' cannot be separate from the contexts which give rise to them. In education, the appeal is not for pedagogies of fun, therapy, or to 'be happy', but for approaches that change the very notion of happiness itself. In this way, we insist that happiness is an important topic for policymaking and research, but are not naïve to its potential pitfalls. Moreover, to avoid any misunderstanding we would like to state clearly that the goal of developing happiness and well-being indicators is not to control or manipulate emotions, much less to rank or create competition around well-being. Instead, the goal is to raise awareness of the contexts and conditions that support happiness, so that these different contexts can be enriched; to initiate a richer, wider debate than we have had to date about the values and requirements to get there within a given society.

## Global Happiness Indices: How to Compare?

We have seen above the inevitable tension between objective and subjective indicators, on the one hand, and continued opacity around the link between micro well-being measures and macro-level units such a countries and/or regions, on the other. How can we understand the level of happiness for an entire country? Is it possible to compare across countries? If so, do we gain anything meaningful by doing so? These questions persist, despite the dramatic rise of global happiness comparative indices over the past decade. Each of the different indices we reviewed in Chap. 1 has approached these questions in different ways. In this section, we will look closer at a handful of the most prominent of them—the Social Progress Index, World Happiness Report, and Better Lives Index—with the aim of clarifying how they produce their indicators and rankings. This is important as we contemplate, largely in Chap. 5, how future global indices might be changed to better account for differences across different cultural contexts globally.

If we arranged the most prominent global indices along an imaginary continuum from objective to subjective, along the furthest end of the objective side of the scale would be the Social Progress Index (SPI). Similar to the UN's Human Development Index (HDI) that incorporated life expectancy and school attainment with income, the SPI aggregates a wide range of indicators, as shown in Fig. 3.3. The SPI indicators selected can vary slightly by country, depending on available data. One key point to note here with the SPI is the lack of traditional economic data (GDP or GNI). But equally important, in the context of our discussion, is the near total *lack* of subjective indicators in the SPI. The only place where subjective indicators arise is in the item 'Like what I do every day', under the category of Opportunity: Freedom & Choice. This refers

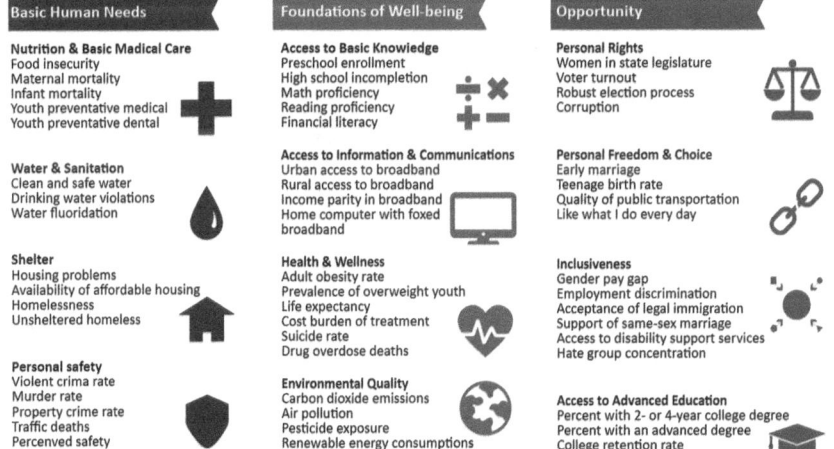

**Fig. 3.3** Social Progress Index items. (United States Version)

to the percentage of respondents who agreed or strongly agreed with the statement 'I like what I do every day' in the Gallup polls. Here it is interesting to note that, again, 'Personal Freedom and Choice' are equated with an individual ('I') being able to do whatever they want. That said, this single subjective item is almost completely overwhelmed by the sheer number of objective indicators that surround it.

So what were the results? These are shown in Fig. 3.4. Countries are assigned a composite score in each major category, and these are averaged for a total score (Social Progress Score). Countries are then ranked, and placed into six tiers. In the context of our larger discussion, we note that Japan ranks in Tier 1, closely followed by South Korea, and well-ahead of

| SPI Rank | Country | Social Progress Score | Basic Human Needs | Foundations of Wellbeing | Opportunity |
|---|---|---|---|---|---|
| 1 | Norway | 92.63 | 95.29 | 93.30 | 89.30 |
| 2 | Finland | 92.26 | 95.62 | 93.09 | 88.07 |
| 3 | Denmark | 92.15 | 95.30 | 92.74 | 88.41 |
| 4 | Iceland | 91.87 | 96.66 | 93.65 | 85.04 |
| 5 | Switzerland | 91.78 | 95.25 | 93.80 | 85.28 |
| 6 | Canada | 91.41 | 94.72 | 92.83 | 86.68 |
| 7 | Sweden | 91.20 | 95.50 | 91.85 | 87.26 |
| 8 | Netherlands | 90.57 | 94.09 | 91.24 | 86.38 |
| 9 | Japan | 90.44 | 96.85 | 92.93 | 81.53 |
| 10 | Germany | 90.32 | 94.59 | 89.35 | 87.03 |
| 11 | Australia | 90.28 | 95.12 | 90.41 | 85.33 |
| 12 | New Zealand | 90.02 | 94.30 | 91.50 | 84.27 |
| 13 | Ireland | 89.47 | 93.91 | 89.58 | 84.93 |
| 14 | Austria | 89.44 | 94.42 | 91.32 | 82.58 |

Fig. 3.4  SPI ranking (selected countries)

the UK and USA. Japan is also ahead of France, Belgium, Spain, and Italy. Moreover, in the 'Foundations of Well-Being' category, Japan ranks fourth in the world, ahead of Denmark, Sweden, Canada, the Netherlands, and Germany, among others.

Next let us look at the World Happiness Report (WHR). Returning to our imagined continuum of global indices, the WHR would be located on the furthest end of the subjective side. In 2021, the WHR collected data on and ranked 119 countries, largely in partnership with the Gallup organization. The WHR utilizes the Cantril Self-Anchoring Ladder we discussed earlier, asking respondents to imagine their best possible life and then rate their current lives on a scale from 0 to 10. Given the disruptions of the COVID pandemic, the 2022 WHR took an average of three years, a period spanning 2019–2021. In the latest reports, the WHR has attempted to expand upon the Cantril Ladder by including frequency of positive and negative emotional states. This calculation is complicated, so we won't go into it here. The WHR report also collects objective indicators, but only uses these to estimate which factors contribute to the life evaluation scores. The indicators are grouped into six factor domains: levels of GDP, life expectancy, generosity, social support, freedom, and corruption. The results are shown in Fig. 3.5. As mentioned in Chap. 1, Finland has been ranked the 'happiest' country for five consecutive years. Finland is followed almost exclusively by northern European countries: Denmark, Iceland, Switzerland, Netherlands, Canada, Sweden, Norway, and so on. Japan ranks 54, and South Korea 59.

Last, let us look at the OECD's Better Lives Initiative (BLI). In many ways, the BLI represents a midpoint on our imagined continuum. Launched by the OECD in 2011, it coincides with the UN's *Happiness* resolution and the move away from GDP led by economists Sen and Stiglitz. As we touched upon in a previous chapter, the 11 topics that comprise the BLI consists mostly of objective indicators (e.g., income, jobs, housing, health, and education). But it does include one major subjective indicator, the now familiar subjective well-being (life satisfaction), as measured by the Cantril Ladder. Figure 3.6 summarizes what the BLI measures. Despite the overwhelming focus on objective indicators, what places BLI in the midpoint is the ability for users to weight the different topics according to their preference. "What is your recipe for a better

**Rank Country**

| Rank | Country | Value |
|------|---------|-------|
| 1. | Finland | 7.821 |
| 2. | Denmark | 7.636 |
| 3. | Iceland | 7.557 |
| 4. | Switzerland | 7.512 |
| 5. | Netherlands | 7.415 |
| 6. | Luxembourg | 7.404 |
| 7. | Sweden | 7.384 |
| 8. | Norway | 7.365 |
| 9. | Israel | 7.364 |
| 10. | New Zealand | 7.200 |
| 51. | Hungary | 6.086 |
| 52. | Mauritius | 6.071 |
| 53. | Uzbekistan | 6.063 |
| 54. | Japan | 6.039 |
| 55. | Honduras | 6.022 |
| 56. | Portugal | 6.016 |
| 57. | Argentina | 5.967 |
| 58. | Greece | 5.948 |
| 59. | South Korea | 5.948 |
| 60. | Philippines | 5.904 |
| 61. | Thailand | 5.891 |

**Fig. 3.5** WHR rankings, 2019–2021 (selected countries)

life—a good education, clean air, nice home, or money?" asks the BLI official website, as it welcomes users to "create your own Better Life Index" by weighting each topic differently. We encourage our readers to put down this book for a moment and take a look at the BLI website. There users may adjust the relative importance of a given variable to better 'fit' their context. In other words, subjective evaluations appear both in the measurement (life satisfaction) and in the weighting/reporting (arrangement of the BLI). This is rather innovative and welcome, although it is not without problems, as we shall see later. The OECD also catalogs the responses it gets from users, claiming upward of 100,000 cataloged responses to date. Tellingly, it shows differences in the relative weighting

| Dimension | Indicator | Unit |
|---|---|---|
| Housing | Dwellings without basic facilities (−) | Percentage |
| | Housing expenditure (−) | Percentage |
| | Rooms per person (+) | Ratio |
| Income | Household net adjusted disposable income (+) | US Dollar |
| | Household net financial wealth (+) | US Dollar |
| Jobs | Labour market insecurity (−) | Percentage |
| | Employment rate (+) | Percentage |
| | Long-term unemployment rate (−) | Percentage |
| | Personal earnings (+) | US Dollar |
| Community | Quality of support network (+) | Percentage |
| Education | Educational attainment (+) | Percentage |
| | Student skills (+) | Average score |
| | Years in education (+) | Years |
| Environment | Air pollution (−) | $\mu g/m^3$ |
| | Water quality (+) | Percentage |
| Civic engagement | Stakeholder engagement for developing regulations (+) | Average score |
| | Voter turnout (+) | Percentage |
| Health | Life expectancy (+) | Years |
| | Self-reported health (+) | Percentage |
| Life satisfaction | Life satisfaction (+) | Average score |
| Safety | Assault rate (−) | Percentage |
| | Feeling safe walking alone at night (+) | Percentage |
| | Homicide rate (−) | Ratio |
| Work-life balance | Employees working very long hours (−) | Percentage |
| | Time devoted to leisure and personal care (+) | Hours |
| | Employment rate of women with children (+) | Percentage |

**Fig. 3.6**   BLI indicators (as summarized in Mehdi, 2019)

given by different contexts. For example, in the USA, UK, Germany, Finland, Sweden, and the Netherlands, most users placed the emphasis on Life Satisfaction. In France and Spain, the emphasis was on Health. Meanwhile, users in Japan selected Safety as their most important concern (OECD, 2022).

It is important to note here that the BLI's user-led weightings are still restricted to what is measured objectively. For example, the BLI Education score is a composite of 'educational attainment' (the number of adults 25–64 holding at least an upper secondary diploma), 'student skills' (average PISA scores of 15-year olds across all three domains: reading, mathematics, and science), and 'years in education' (average duration of education between 5 and 39 years old), which attempts to get at how many people continue past secondary education. Or to take another example, the BLI Environment score is composed of 'air pollution' (population weighted average of annual concentrations of 2.5 PMI) and 'water quality' (captured by percentage of 'satisfied' responses to the question "in the city or area where you live, are you satisfied or dissatisfied with the quality of the water"). Thus, when users 'weight' Education as their top priority, they are doing so subjectively but within a set of objective parameters set by the OECD. In the case of Environment, for example, items like $CO_2$ emissions per capita that would relate Environment to climate change are not included. Some OECD reports attempt to correlate the objective measures to Life Satisfaction, with one OECD analysis purportedly showing income and housing have the highest correlations with the overall Life Satisfaction of a given country. And despite the OECD appeal to users to create their own balance of measures and its clear attempt to avoid rankings, in some places—including the main BLI website—the OECD calculates the 11 topics equally to derive a ranking, as shown in Fig. 3.7. For 2022, the first eight slots were taken by Norway, Iceland, Switzerland, Sweden, Finland, Netherlands, Australia, and the United States. Japan ranked 30, and South Korea 32, among the roughly 40 OECD countries included.

## National/Local Happiness Indices: Similar Issues?

Although our main focus in this volume is the global context, it is worth briefly mentioning how happiness and well-being is debated and operationalized in national and local policymaking arenas. This helps bring into further relief the tensions and debates that are intrinsic in this sort of work, wherever it may be carried out. Here we focus on the Japanese government's Ad Hoc *Research Council on Well-Being*, inaugurated in 2011. The Council was charged with creating the first ever well-being indicators for Japan. One of the current authors (Uchida) was selected as a member of the expert group, which included economists, sociologists, experts in public policy, and opinion leaders from various fields. It was that experience that gave us first-hand knowledge of how contentious, difficult, and—in a sense—arbitrary the process of creating indicators can be. It also shed light on the interplay of global rankings and national/local policy processes.

**Fig. 3.7** OECD's BLI Index, default rankings. (OECD, 2022)

In the course of the multiple deliberations of the Council, the following points of concern and contention emerged:

- Pros and cons of indexing subjectively experienced happiness versus objective well-being, and examining the myriad factors that influence an overall sense of happiness.
- Proper balance between subjective and objective indicators.
- Whether or not international comparisons conducted by the OECD and other international organizations were even useful, in the sense of being inclusive of variations in well-being found in the Japanese context.
- Whether or not individual evaluations and/or subjective well-being indicators could be utilized effectively in more specific policy domains, such as social security, medical care, aging policy, and education (child-rearing), or whether, to be useful at all, these measures had to capture something more expansive in scope.
- In light of Japan's aging population and vast changes between generations (i.e., from postwar destruction/rebuilding to high-economic growth to two decades of recession), what sorts of surveys or measurements could simultaneously take into account life stages and generational differences?
- Would it be appropriate to measure only individual happiness? Or would it be more appropriate to measure at the level of households, local communities, and regions, to recognize disparities between groups, thus making targeted policymaking more feasible?
- Should results be 'ranked' according to Japan's 47 prefectures? Should the results be reported as an easy-to-understand single numerical value ('Happiness Index')? Or, considering how multifaceted the nature of well-being is and the negative impacts of simple ordinal ranking, should results be reported in alphabetical order and detailed in full complexity?

In the end, the Council's Final Report submitted in December 2011 not only featured no composite Happiness Index or prefectural ranking, but actually carried an explicit and stern warning against creating rankings: "Regarding the use of an composite index, since the purpose of the happiness index is to identify the positive and negative points of Japanese society and to consider how to tackle those issues, the creating of an composite index would, on the contrary, lead to the obfuscation of the specific characteristics of each domain" (Cabinet Office, Government of Japan, 2011, p. 32). That is, the Council came to the conclusion that even though composite indictors (i.e., 'rankings') grab public attention, such forms of reporting too often hide nuances within the measurement and create distortions, simply due to arbitrary methods of weighting and subsequent integration.

More noteworthy than the results was the conceptual scheme developed. 'A Feeling of Well-Being' was set as the overarching goal, but this was understood in a multifaceted way. As shown in Fig. 3.8, it included subjective well-being, positive/negative emotional states, and well-being with others. We delve deeply into some of the elements of this conceptualization in the next chapter. Underneath this were three supporting pillars: 'physical and mental health', 'socio-economic conditions' such as work and living environment, and 'relationality' such as connections

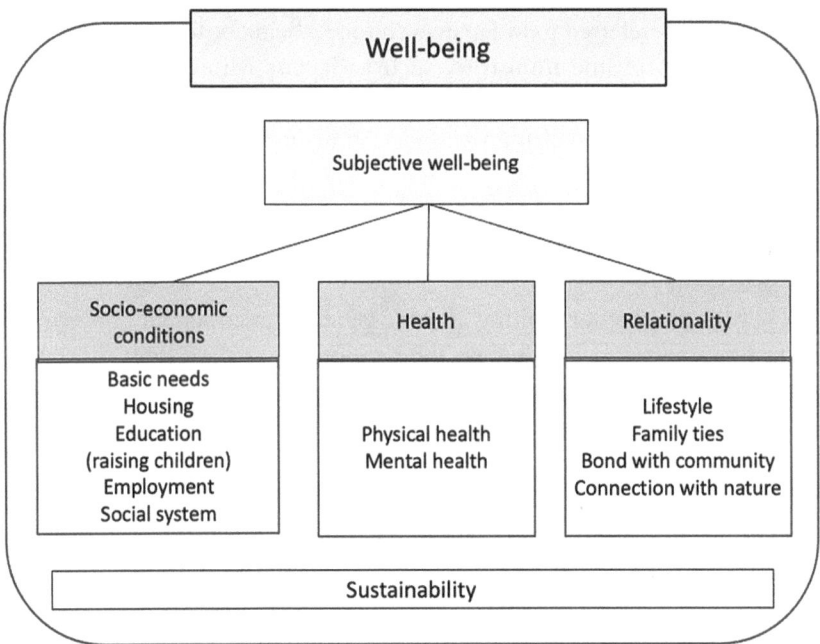

**Fig. 3.8** Japan's Research Council on Well-Being, overarching conceptualization (2011)

with family, community, and nature. Particularly this last item, 'connection with nature' may be particularly innovative in the global context. Note that the Council conceptualized sustainability—in the sense of social and environmental dimensions—along a separate axis. Based on this conceptualization, the Council fielded proposals among the wider public (academics, business leaders, civil society groups) for specific indicators, both subjective and objective, that could be used to measure each of these components. In the end, more than 130 proposals were submitted, giving a glimpse of the diversity—even within Japan—about how values and indicators can be paired.

Unfortunately, Japan's Well-Being Council was disbanded due to political changes. This left many of the discussions unresolved, and the proposed indicators were not adopted. This underscores how, as discussions around happiness and well-being become more politically mainstream, the less stable they can become, as compared with the long-term commitment of researchers. That said, the draft indictors and Council discussion materials are continually referred to in Japan even today, being utilized by some local governments and line ministries, such as Japan's Ministry of Education (MEXT) which, under the influence of the OECD and PISA, are now interested in operationalizing well-being in its specific policy domain.

## Conclusion: (Mis) Measurement

We began this chapter looking at some general directions for conceptualizing happiness and well-being, then moved to outline how those tensions are found in the measurement choices of the most prominent global indices. We then turned to a discussion of actual policymaking deliberations around well-being in the Japanese context. Our purpose was to bring to the fore questions about both how to measure and why we seek to measure. The key point is that any measure is a mismeasure as well: any measure both elucidates and obfuscates/distorts. This is all the more true at the global level, where there is no consensus on the 'best' way to measure, given differences in how different cultures live. To unpack this a bit more, we bring this chapter to a conclusion by sketching some of the problematic dimensions of the contemporary ranking-policymaking excercises, both those seen and unseen.

The use of simple indicators, although they gain public attention, often tends to narrow discussions, leading quickly on to efforts to 'boost' one's own rank or 'overtake' a perceived rival. That which is too easy to understand often gets over utilized, rather than becoming a catalyst for deeper discussion and analysis. This, again, is why Japan's Well-Being Council elected not to rank—a level of caution about potential policy and popular effects we do not currently see at the global level. Part of the problem here is the use of averages. Averages are a necessary evil within rankings across different units, but averages simultaneously hide differences within a given unit. Indicators can be useful for determining strengths and weaknesses of a given set of policies or practices in a given situation, that is, whether or not regional, class, and/or familial gaps in happiness are widening, or whether economic wealth has—over time—given way to, say, environment as the key determinant of happiness.

But rankings—a hierarchical arrangement based on averages—too often lead to competition. Competition and 'solutions' tend to short-circuit richer discussions on the specific 'meaning' of happiness for given groups of people, how these meanings shift over time, and to create context-specific policies based on that more nuanced understanding. This is why it is crucial to be able to disaggregate data, say, between youth, middle-aged, and older people to correlate changes in context with the life-cycle of individuals, or to be able to understand community factors that link to individual subjective well-being. Part and parcel to this are surveys conducted over time: panel data allows us to see broad term temporal changes, and how changes in social policy may affect those outcomes, as opposed to only comparing against other units (i.e., countries) at a single point in time. Unfortunately, many of the global indices are not panel data. For example, the OECD asks different questions on different years which makes comparison difficult, something we discuss further in Chap. 5. Overall, our message here is to caution against narrowly measuring happiness with an eye toward simplified rankings. We need to instead encourage creation of a richer, more nuanced conversation by including a wider range of voices. Through this process, the diversity of meanings assigned to the polysemantic term 'happiness' and 'well-being' will inevitably emerge. Only within that richer, more nuanced conversation can specific policy interventions have any meaning, and any chance of success.

This point leads right into the next two chapters. In discussions at the global level, we must recognize that the same dynamics are at play. In fact, the dynamics are more fraught. We may welcome the collective shift toward happiness and recognize the need for some commonality of shared indicators, but we have to insist on the 'uniqueness' of different countries, regions, and cultures that constitute the 'global'. If the starting point of conceptualizing happiness is different, translating and using indicators formulated in one country or region is unlikely to be effective. If response styles differ based on culture and/or our reference group when making self-evaluations, then seemingly common measurements actually *mismeasure*, distorting rather than clarifying images of the world.

Concretely, in this chapter we have seen how the 'rank' of Japan varies widely according to which global survey is utilized: from a Tier 1 (#10) showing in the SPI, to a middling showing in the OECD's BLI (#32), to a dismal showing in the WHR (#54). This underscores that much more is at play in measurements and indicators than objective reality alone. Some scholars are so insistent that global measures are actually (mis) measures that they want to completely do away with the rankings. Yet, we are of the opinion that carefully recognizing what is being measured, how different measures reveal different things, and earnestly examining why differences emerge, are useful in the pursuit of a larger goal: understanding the differences in the meaning, structure, and content of happiness across *different* countries and cultures. Rather than rushing to rank, we can use the measurements in service of diversifying theory, policy, and practice alike.

# References

Adams, G., Estrada-Villalta, S., Sullivan, D., & Markus, H. (2019). The Psychology of Neo-Liberalism and the Neoliberalism of Psychology. *Journal of Social Issues, 75*, 189–216.

Cantril, H. (1965). *The Pattern of Human Concerns*. Rutgers University Press.

Diener, E., & Biwas-Diener, R. (2008). *Happiness: Unlocking the Mysteries of Psychological Wealth*. Blackwell.

Diener, E., Emmons, R. A., Larsen, R. J., & Griffin, S. (1985). The Satisfaction with Lifescale. *Journal of Personality Assessment, 49*, 71–75.

Diener, E., Inglehart, R., & Tay, L. (2013). Theory and Validity of Life Satisfaction Scales. *Social Indicators Research, 112*, 497–527.

Frey, B. (2018). *Economics of Happiness*. Palgrave Macmillan.

Gallup. (2021). Understanding How Gallup Uses the Cantril Scale. https://news.gallup.com/poll/122453/understanding-gallup-uses-cantril-scale.aspx

Helliwell, J. F., & Putnam, R. D. (2004). The Social Context of Well-Being. *Philosophical Transactions of the Royal Society of London. Series B: Biological Sciences, 359*(1449), 1435–1446.

Huta, V., & Waterman, A. (2014). Eudaimonia and Its Distinction from Hedonia: Developing a Classification and Terminology for Understanding Conceptual and Operational Definitions. *Journal of Happiness Studies, 15*, 1425–1456.

Cabinet Office, Government of Japan. (2011). Measuring National Well-Being—Proposed Well-being Indicators. Available in English Online at: https://www5.cao.go.jp/keizai2/koufukudo/pdf/koufukudosian_english.pdf

Kuppens, P., Realo, A., & Diener, E. (2008). The Role of Positive and Negative Emotions in Life Satisfaction Judgment across Nations. *Journal of Personality and Social Psychology, 95*, 66–75.

Madsen, O. J. (2014). *The Therapeutic Turn: How Psychology Altered Western Culture*. Routledge.

Matson, E. (2017). *Perspectives from Smith on Wealth and Happiness, in Adam Smith Works* (Liberty Fund, John Templeton Foundation). https://www.adamsmithworks.org/documents/perspectives-from-smith-on-wealth-and-happiness

Mehdi, T. (2019). Stochastic Dominance Approach to the OECD's Better Life Index. *Social Indicators Research, 143*, 917–954.

Nagel, T. (1972). Aristotle on Eudiamonia. *Phronesis, 17*, 252–259.

OECD. 2018. *What Matters the Most to People? Evidence from the OECD's Better Life Index Users's Responses*. OECD. https://www.oecd.org/officialdocuments/publicdisplaydocumentpdf/?cote=SDD/DOC&docLanguage=EN#:~:text=The%20OECD%20Better%20Life%20Index%20is%20an%20interactive%20composite%20index,users%20since%202011%20to%20date

OECD. (2022). *Create Your Own Better Life Index*. OECD. https://www.oecd-betterlifeindex.org/#/11111111111

Oishi, S. (2002). The Experiencing and Remembering of Well-being: A Cross-cultural Analysis. *Personality and Social Psychology Bulletin, 28*, 1398–1406.

Oishi, S. (2009). Shiawase wo kagasuru: shinrigaku kara wakatta koto [*Towards a Science of Happiness: What Psychology Teaches Us*]. Shinyosha.

Oishi, S., Kesebir, S., & Diener, E. (2011). Income Inequality and Happiness. *Psychological Science, 22*, 1095–1100.

Oishi, S., Schimmack, U., & Diener, E. (2012). Progressive Taxation and the Subjective Well-Being of Nations. *Psychological Science, 23*, 86–92.

Pavot, W., & Diener, E. (1993). Review of the Satisfaction with Life Scale. *Psychological Assessment, 5,* 164–172.

Ram, R. (2010). Social Capital and Happiness: Additional Cross-Country Evidence. *Journal of Happiness Studies, 11,* 409–418.

Vazire, S. (2006). Informant Reports: A Cheap, Fast, and Easy Method for Personality Assessment. *Journal of Research in Personality, 40,* 372–481.

Veenhoven, R. (2010). Greater Happiness for a Greater Number: Is That Possible or Desirable? *Journal of Happiness Studies, 11,* 605–629.

Vintimilla, C. (2014). Neoliberal Fun and Happiness in Early Childhood Education. *Journal of Childhood Studies, 39,* 79–87.

White. (2007). A Global Projection of Subjective Well-Being: A Challenge to Positive Psychology. *Psychtalk, 56,* 17–20.

# 4

# Culture and Happiness: An Interdependent Approach

Previous chapters have progressively prepared the way for the core contribution of the current volume: the explication of an interdependent mode of happiness and well-being. In the current chapter, we examine interdependent happiness in detail, exploring its relation to culture, looking at what it means, what causes it, and the way it shapes motivation. In the course of this discussion, we necessarily discuss the definition of culture and the relationship between, say, culture and environment and biology. We contrast this interdependent mode with the independent mode that, in many ways, has become the default mode for global measurements of happiness. To further bring the interdependent mode into relief, we summarize several key studies and other empirical data developed within the field of cultural psychology. The current chapter sets the stage, in turn, for the next chapter in which we explore how this alternative mode of happiness leads to clear differences in actual cultural practices (e.g., education), social organization, and approaches to measurement.

## Defining Culture, Conducting Comparisons

One dominant theme thus far has been the complexity of understanding happiness and well-being. Some seek victory, accolades, and monetary rewards, while others seek a peaceful, comfortable, and safe life. We can

© The Author(s) 2024

Y. Uchida, J. Rappleye, *An Interdependent Approach to Happiness and Well-Being*, https://doi.org/10.1007/978-3-031-26260-9_4

recognize such variations within any given context, worldwide. Looking back historically, we can find such variations as well: from, say, the comedies of Aristophanes to the tragedies of Aeschylus in the Ancient Greek context, as well as in the differences among the three great philosophical traditions of ancient China. In the famous portrait of the Vinegar Tasters, three men representing Confucius, Buddha, and Laozi are all dipping their fingers in the same vat of vinegar to get a taste. One reacts with a sour expression, another with a bitter one, and another with a sweet one. These different reactions suggest the differences in approach to life and happiness in these different traditions: a Confucian rule-based morality to correct sour behavior; a Buddhist approach of non-attachment to the bitterness of pain, suffering, and life's impermanence; and a Daoist realization that anything in its natural state is neither good nor bad but depends on how we look at it. Orientations toward happiness also change depending on various life circumstances and life stages of the individual, as when one learns to view what once looked bitter as actually sweet.

However, happiness and well-being do not come solely down to individual preferences, orientations, or outlooks. There undoubtedly exist common prerequisites for happiness. Indeed, modern societies have developed social systems (institutions)—for example, education, welfare, public healthcare, public order—that seek to guarantee happiness by promoting a set of common elements, although as we know the actual achievement of those systems varies widely. Moreover, as we have seen, recognition of common patterns has led to scholarly conceptualizations of happiness, including Maslow's Hierarchy of Needs. That is, happiness and well-being derive both from individual personality/orientations/outlooks on life, but also according to the social and cultural circumstances in which 'individuals' are embedded. In the same way that individuals are not uniform in their outlooks, it is also true that the social and cultural contexts in which they are embedded are also far from uniform. Even now when most of us live in modern societies it is clear that collective differences endure. These differences reach all the way back, to return to our example above, to the Greeks who watched theatrical tragedy to make sense of life and to ancient Chinese who debated the 'flavor' of reality but through entirely different metaphorical schemes.

Our focus here is these different social and cultural circumstances. This focus is not intended to deny individual differences: certainly one can live in a cultural context where a given approach to happiness dominates and yet find one's own happiness in a very different way. Here we might think of the individual choice of many Westerners to practice yoga and meditation. Instead, our focus on the social and cultural dimension is an attempt to bring balance back to these discussions. It is now rather common—particularly in the Anglophone world, especially in work emanating from the United States—to view happiness as entirely a matter of *individual* choice and outlook. Yet, it is clear that both individual variation and wider sociocultural patterns shape happiness worldwide. The impact of socio-cultural circumstances becomes most clear when one invokes comparisons across social and cultural contexts. That is, only when research itself is embedded in context does the impact of a given context become evident.

The problem with so much of the research on happiness in the twentieth century is that it lacked comparisons across cultures. This has led to the unfortunate conclusion that findings from one culture explain what is happening everywhere. In fact, most of the dominant psychological theories to date have been developed by North American researchers based on the results of studies conducted on American college students, an unfortunate situation illustrated in Fig. 4.1. That is, research findings derived from a small percentage of Americans (less than 5% of the global population), who are usually of upper and middle-class background (that is, those who were able to, encouraged, and could afford to go to college), are being unthinkingly applied to people living in myriad contexts worldwide. Henrich et al. (2010) utilize the comical but pointed term WEIRD (Western, Educated, Industrialized, Rich, Democratic) as shorthand for these subjects of previous psychological studies. The homogenous background of these WEIRD people extends beyond mere class and educational categories as well: their religion, language, and cultural experiences all differ from the 'rest' of the world.

Of course, some of this WEIRD work is important: it has helped capture universal human psychological tendencies. However, there are increasingly strong warnings—much of it driven by the rise of cultural psychology and comparative social science—that an overly universalistic

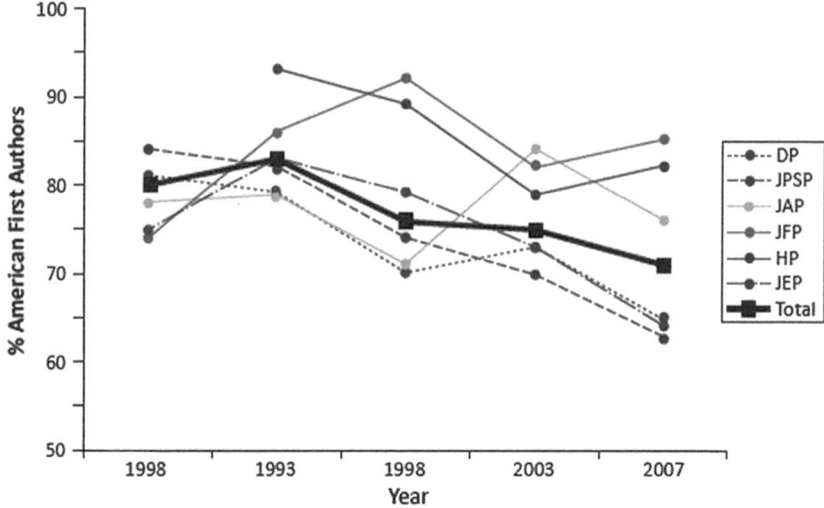

**Fig. 4.1** Percentage of articles published in top psychology journals from 1988 to 2007 with an American as the first author (Arnett, 2008). Abbreviations to the right refer to leading journals in the field of psychology: *DP* Developmental Psychology, *JPSP* Journal of Personality and Social Psychology, *JAP* Journal of Abnormal Psychology, *JFP* Journal of Family Psychology, *HP* Health Psychology, *JEP* Journal of Educational Psychology

WEIRD lens ultimately marginalizes diversity and tries to understand the world without any mention of 'culture'. Indeed, our own research and the current volume have developed from self-awareness that the world is not WEIRD.

But before delving further into this, we need a working definition of 'culture'. Culture in psychology can be usefully defined as "a pattern of values, thoughts, and reactions that has been developed and shared by a group of people throughout their social and group history" (Kitayama, 1998, see also Adams & Markus, 2004). In fact, each of us lives in a wide range of 'societies', from small groups such as the family and friends to larger groups such as regions and countries. In a given society, people interact with each other, certain behavioral habits and norms are created, and values are transmitted among members within a given generation and, in a subsequent step, from one generation to the next (e.g., from parent to child; via education and schooling). 'Culture' is the tangible and intangible frameworks that are shared within a given society, a composite we often refer to as customs, norms, and values. As culture is shared and transmitted by a group of people, and as the people who make up a given group change, cultures inevitably change. Cultural psychologists sometimes use the term 'group' or 'subgroup' to capture ethnic or gender groups within a larger political entity that may not necessarily share a single culture. They also use the term 'group' to signify any body of people who share a historically accumulated, shared background, such as a region or a family.

Interestingly, culture implies habits and rules of conduct that are so central to who we are that often we are not explicitly aware of them. We simply take for granted how we greet people, express emotions, attempt to get along with family and intimate others, and what it means to help other people. We also take for granted how we think about a problem, how we express disagreement, and forge consensus. But these are, in nearly all cases, expressions of culture. Since these customs, rules, values, and other cultural aspects of the group to which we belong are shared by the people around us, and are so deeply integrated into our daily lives, we are rarely aware of them. They appear as simply 'natural', even to the point that some people deny culture itself exists. But this sort of a-cultural view quickly breaks down the moment one comes into contact with a group that shares different norms and customs. In a globalizing world, this is happening with increasing frequency. Readers may have experienced this when visiting a foreign country and becoming surprised—or troubled—at just how different the ways of life and/or seemingly universal values really are (e.g., what and how we eat; how we think about family; how we express love).

Here it is worth engaging some of the recent debates surrounding 'culture', particularly as we envisage our readers come from a variety of backgrounds. The field of cultural psychology, whose findings comprise the foundations of this volume, originally arose from the merging of insights from psychology and anthropology in North America. Anthropologists, particularly those in North America, were focused on understanding how particularly groups (e.g., Navajo in the American Southwest) made sense of the world and how different cultures understood and experienced the world differently (Kroeber & Kluckhohn, 1952/1966). Meanwhile, the field of psychology, which had developed largely from Western philosophical categories (e.g., cognition, emotion, and motivation; Reason), unfolded under a premise of universalism: that everyone worldwide shared the same basic psychological structure and experience. In the

1960s–1970s, anthropology and psychology were brought into dialogue, largely through efforts initiated by anthropologists. Richard Shweder's work at the University of Chicago was pivotal, as elaborated in works like *Culture Theory: Essays on Mind, Self, and Emotion* (1984). This transdisciplinary dialogue led to a flood of work in the late 1980s–1990s that utilized anthropological insights to test and challenge WEIRD psychology categories and theory. Catherine Lutz's *Unnatural Emotions: Everyday Sentiments on a Micronesian Atoll and Their Challenges to Western Theory* (1988) is a representative and powerful example.

However, just as cultural psychology was taking off, anthropologists themselves were turning against the term 'culture'. Originating in the philosophical currents of deconstructionism and poststructuralism in the 1980s–1990s, 'culture' was seen as distorting for any number of reasons: it projects an image of coherence and homogeneity that does not exist, it is too abstract, it suggests consensus but obfuscates power, it suggests an enduring essence of a given people, it is too easily manipulated for nefarious political and racial ends, and so on. So strong were these critiques that by the late 1990s many in the mainstream social sciences became averse to using the term 'culture' altogether. On the one hand, mainstream social science disciplines like sociology had not expanded beyond Western assumptions, while, on the other hand, to label something as 'cultural'— that is, *cultural* psychology—was to invite the easy critique that one has not engaged with the latest self-reflections in philosophy and anthropology. Thus, from the vantage point of contemporary social science, it is too easy to portray cultural psychology as outdated and out-of-touch.

Obviously we do not share such a view. Not only have the social sciences started to understand their narrowness (via post-colonial and decolonial movements), but culture is returning to recent discussions. Our own view is highly pragmatic: we need a way to speak about these different frameworks of making sense of the world, and retaining the word 'culture' is one way to do so. Perhaps another term might work. Earlier in this volume we used the term "cognitive frames"; in other work we use "worldview" or "ontological" (Rappleye, 2020; Rappleye & Komatsu, 2017). Yet, arguably nothing works quite as well as 'culture'. We are thus in agreement with Brumann (1999) who argues that, despite the fact that there is great diversity in our global human landscape, it is constantly changing, relationally co-created, and not always consensually arising, "we still need a vocabulary for describing its mountains, plains, rivers, oceans, and islands. The anthropological concept of culture offers itself for that task" (S13). That is, the concept of culture is pragmatically productive for the task of diversifying the larger global cartography around the ways people think and what makes them happy.

For us then, utilizing the term culture always means culture*s*, and implies comparison. Indeed, the only means of avoiding an essentialist cultural argument is to understand a given culture in relation to another; to view it in contrast to a different pattern of meaning and experience of the world. In this volume, our point of comparison is the dominant patterns of North America. Not only are the North American patterns dominant in contemporary scholarship, but the North American patterns are also dominant in the minds of policymakers globally. This is a result of the heavy influence of North American scholarship on policymaking. This fact is evident in the various movements (e.g., Positive Psychology) and indices (e.g., the SPI) currently being produced to measure happiness (Hendriks et al., 2019). As we saw in the last chapter, the subjective scale of life satisfaction carries an implicit assumption of individual, acquisitive forms of happiness. The promotion of these sorts of indicators, as opposed to, say, indicators on social safety, is also indicative. We will come back to explore these links more in the next chapter. Our point here is simply that comparisons with North America are insightful both academically and politically. Moreover, for those of us raised and/or working in East Asia (Japan, in our case), the difference is felt profoundly at an experiential level.

This raises our final point of preliminary consideration: To what degree is the interdependent happiness model we outline below synonymous with Japan, and Japan alone? This is a crucial question for the ambitions of our entire volume. To answer it, let us raise a contrast with the case of Indigenous Psychology to make our own position clear. As psychology began to diversify away from a universalistic model premised on the WEIRD societies of the West, one project that arose was Indigenous Psychology. That school of research focused on the particularly psychological constitution of, say, the Chinese. Indigenous terms like 'face', *guanxi*, and filial piety became the focus. Some of the work sought to link this indigenous psychology to the philosophy of Confucianism. Certainly such psychological phenomena exist and such historical links are evident. Undoubtedly, this school has produced a number of important studies. Nevertheless, there is a certain element of relativism inherent in that work, we feel. That is, it is unclear what the relevance is for non-Chinese. Indigenous Psychology can sometimes give the impression that each cultural grouping has its own psychology, and the work of research is limited to empirical elucidation of these unique psychological landscapes.

In contrast, cultural psychology, while also committed to diversifying mainstream WEIRD psychology, focuses on what is quasi-universal. This means that people everywhere are capable of—*and do!*—experience psychological states that are different from the mainstream WEIRD view. In the relations between mother and child, the coordinated movements of a sports team, and the feel of calm 'well-being' following yoga or a walk in the forest, we may recognize psychological states that are different from mainstream psychological views, but in no sense limited to, say, the Japanese, Chinese, or Indians. Indeed, the global spread and enthusiastic reception of non-Western practices of well-being such as yoga and meditation (mindfulness) suggest that these psychological states are shared and valued widely, but still unrecognized by mainstream WEIRD scholarship. It is within a 'middle' space that cultural psychology and this volume work, and seeks to contribute: eschewing local relativism and a WEIRD 'one world' scenario, in favor of a diversification of our understandings of what we may, in fact, all share in common. Hence, the point is not a 'Japanese' model versus an American one, or an 'Eastern' model versus a 'Western' one. Instead, it is the search for shared experiences that are not currently reflected in existing measures, models, theories and/or practices; a pragmatic approach that seeks new ideas on a global scale from diverse cultural repertories.

# Exploring an Interdependent Approach to Happiness and Well-Being

So what then is an interdependent approach? Let us begin in North America. There a 'happy' person is defined, in general terms, as one who is young, healthy, well-educated; has a high income (enabling one to acquire what one desires); is outgoing and optimistic; and—crucially— possesses high self-esteem (Myers & Diener, 1995). Research results have confirmed that, among these factors, being healthy and having high self-esteem are the two factors among these which have the biggest impact on North American happiness. Figure 4.2 schematically concep- tualizes these general differences, looking at feelings invoked when one is 'happy', how being 'happy' is conceptualized, and the predictors of happiness. Being happy in North America is also associated with feeling an elevated stated ('high'), standing out, and a sense of pride. Happiness is understood as a state that potentially increases infinitely, is wholly positive, and, thus, should be pursued to the fullest extent. The domi- nant predictors of happiness in North America become a sense of per- sonal achievement orientation, self-worth, self-respect, and self-esteem. Throughout this volume, we shall call this complex the Independent Happiness approach (or mode).

In contrast, forms of happiness dominant in East Asian contexts, par- ticularly Japan, are different. Not deficient, but different. In East Asia and Japan, we do not observe the same connections between happiness and factors like 'self-esteem', happiness through acquisition ("so far I have got- ten the important things I want in life"), and individual independence.

| | Japan | North America |
|---|---|---|
| Emotions related to Happiness | - Low-arousal emotions (calm)<br>- Relational emotions | - High-arousal emotions (excitement)<br>- Emotions of separation/distance |
| Happiness understood as… | - Reference-based ('with reference to…')<br>- Inclusive of Negative Dimensions | - Expansion Model<br>- Positive |
| Predictors of Happiness | - Relational thinking<br>- Feeling of Fitting-In<br>- Relational Interdependence | - Individual Achievement<br>- Values Self-Determined<br>- Self-Esteem |

**Fig. 4.2** Differences in happiness between Japan and the United States

In fact, when we present such definitions of North American happiness in our classes or public lectures within Japan, the audience often appears confused, and ask simple, but profound questions like 'What does it means that happiness increases infinitely?' These questions and doubts derive from the persistence of different notions of happiness and well-being in Japan: feelings that are more calm and intimate, happiness referenced to relationships, happiness as understood as inclusive of both positive and negative sides. In Japan, the empirically observed, dominant predictors of a happy person are an orientation toward relationships, a sense of fitting in, relational attunement, and social support. All of these qualities constitute the broad outlines of this volume's key terms: Interdependent Happiness and well-being.

These broad classifications derive from several meta-reviews of three decades of existing research on happiness conducted in Japan and North America (Uchida et al., 2004; Uchida & Ogihara, 2012). In surveying this body of research concisely here, we suggest to the reader that it is useful to think in terms of the (1) meaning, (2) motivations, and (3) causes. In other words, to get a clearer picture of the differences between the Interdependent and Independent modes, we may ask (1) what might constitute happiness?, (2) what might people do to achieve happiness?, and (3) what factors might predict happiness? Readers who seek deeper discussion and/or further empirical support may refer to the many academic references we have included at the end of this volume, particularly the review article entitled, *Cultural Constructions of Happiness: Theory and Empirical Evidence* (Uchida et al., 2004). In attempting to keep the current volume widely accessible, both in terms of length and level of specialization, we here offer only a sketch of the broad contours of this enormous body of recent research.

In terms of meaning, it is common to define happiness as simply a positive emotional state. This definition suggests that happiness is universal: feeling good and positive. However, it is crucial to see happiness as embedded in specific contexts and thus open to quite different meanings. We need to move beyond a superficial discourse of happiness as simply 'good' and 'positive' to see the variations in meaning that are at play. In North America, happiness is most typically understood as a state contingent on personal achievement and positivity. In this Independent

Happiness mode, personal achievement confirms one's abilities, leading one to feel a sense of pride and accomplishment. '*He had a deep sense of pride around what he accomplished*' is a common refrains one hears in the North American context, suggesting a feeling of pleasure or satisfaction derived from one's own achievements. The point of refererence—that is, where attention is focused when making this judgment—is individual, internal, and independent. It is independent in the sense that it does not refer to contexts surrounding the proud individual.

In contrast, in East Asia happiness is more often understood as a state contingent on social harmony and a balance among different selves in different matrices of relations. Different selves can be a difficult concept for those assuming self-consistency, rather than a perspectival or socially determined self. The word harmony can sound merely quaint. But the word harmony derives from an analogy of music, wherein the simultaneity and concordance of different notes and instruments produces a pleasing sound. Attunement is another word we might substitute for harmony, and helps us to imagine that happiness is akin to 'being in tune' with one's surroundings. These metaphors helps us imagine how, in this Interdependent Happiness mode, the point of reference are encompassing social relations, located externally, and formed in interdependent relations, which inevitably change.

In terms of what motivations underlie happiness, in the Independent mode people tend to pursue happiness by seeking and confirming the positive internal attributes of the self. This manifests in a number of ways, including a tendency for self-enhancement in causal attribution of success and failure (Kitayama et al., 1995), a preponderance toward self-referential judgments (Heine & Lehman, 1999; Markus & Kitayama, 1991), and a predisposition to accept positive feedback about the self (Heine et al., 2001; Kitayama et al., 1997). Heine et al. (2001) note that, in particular, the tendency toward positivity of the self is constantly reinforced and required for one to become a respectable cultural member of North American societies. This manifests empirically in the finding that Americans are far more likely than Japanese to report experiencing positive emotions (Kitayama & Markus, 2000). Interestingly, this is unlikely to be motivated by any substantive difference in actual emotional experience, but instead largely due—as discussed previously—to the tendency for Americans to remember and emphasize positive emotional experience (Oishi, 2002; Diener & Suh, 2000). Americans are also far more likely to seek to achieve personal (read: Independent) happiness.

In contrast, in the Interdependent mode, we find an underlying motivation to pursue communal or intersubjective forms of happiness. Japanese and many other East Asians tend to seek out and rely on the advice, opinions, and judgments of close others—those such as partners or parents—when pursing happiness. This external focus manifests in being more prone to utilize expectations by close others in organizing one's own behaviors (Iyengar & Lepper, 1999). This mode shows a far greater concern for approval and confirmation of close others, and, in turn, the care and support of these close others are impactful on happiness within an Interdependent mode.

These differences in meaning and motivations lead to different predictors of happiness. In North America factors related to personal happiness are the primary predictors of happiness, while in East Asia the most reliable predictors are the realization of social harmony (attunement). In other words, in North America high levels of happiness are predicted by an individual's levels of personal accomplishment and self-esteem. This manifests in a focus on the perception of one's positive attributes, even in cases where it is illusory or unjustified. This focus leads to and contributes to improved mental states. In the next chapter we will elaborate on this further, in regards to measures like life satisfaction and movements like Positive Psychology and Positive Education. However, in East Asian contexts the significance of self-esteem is questionable. Empirical studies such as Diener and Diener (1995) which investigated the relation between self-esteem and well-being across 31 countries found that self-esteem is more strongly correlated with subjective well-being in individualistic (e.g., European-American) cultures than in collectivistic (e.g., East Asian) cultures. Subsequent empirical studies have confirmed that social harmony more reliably predicts happiness in East Asia, as compared with self-esteem (Endo, 1995; Kitayama & Markus, 2000). Consistent with this, Suh et al. (1998) found that positive affect enhances the feelings of happiness in North America, while in East Asia social factors (such as adapting to social norms and fulfilling relational obligations) increase happiness.

More recent studies have simultaneously examined both self-esteem and social harmony in predicting happiness. In one such study (Oishi & Diener, 2001), consideration was given to cultural differences in motivation for goal attainment. Participants in the experiment listed five important goals for the coming month and rated how personal each goal was. One month later, they rated their satisfaction with their lives during the month, and also indicated how well they had achieved the goals they had described a month earlier. As shown in Fig. 4.3, the results showed that while the achievement of Independent goals (obtaining one's own pleasure and enjoyment) had an effect on happiness in European Americans, no such effect was found among Asian Americans. Instead, the pursuit of Interdependent goals (pleasing one's parents or guardians) had an effect on happiness. The results are consistent with other findings on motivation: Asian Americans tend to perform better when given a problem by others, as compared with facing a problem of their own choosing, in contrast to European-Americans who perform better when given a problem of their own choosing (Iyengar & Lepper, 1999).

Overall then, in East Asian cultures, where the Interdependent mode tends to be foregrounded in both the "cognitive frame" and institutions, happines is not predicated on the achievement of individual goals, but

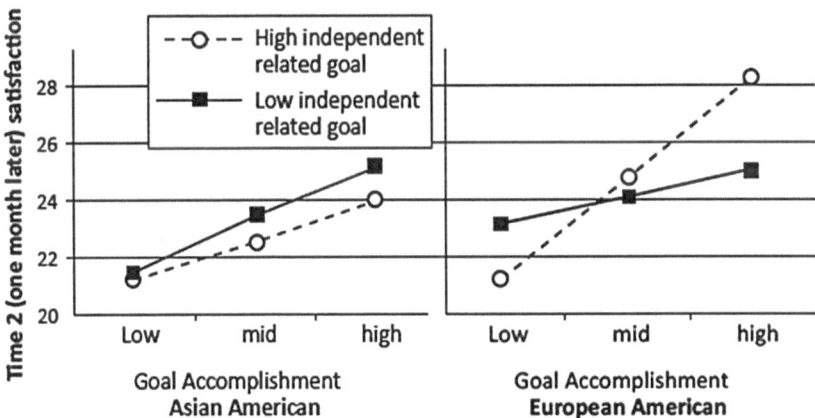

Fig. 4.3   Satisfaction as a function of goal attainment for Asian Americans and European Americans with high and low goal pursuit. (Adapted from Oishi & Diener, 2001)

rather by emotional ties with others (Endo, 1995; Kitayama & Markus, 2000; Triandis, 1995; Suh et al., 1998). The media of attunement is the affective bond. Balance with one's surroundings is key to achieving happiness (Suh, 2002). We rarely see in China and Korea—as with Japan—a strong motivation to prominently display "personal happiness", but rather an emphasis on maintaining a balance within one's surroundings (Suh, 2002). In these contexts, social factors (conformity to value norms) affect well-being (Suh et al., 1998), and this creates a strong 'other' focus perspective on happiness. In the North American Independent mode, where individual characteristics are emphasized, individual emotional experiences affect subjective well-being, happiness is achieved by maximizing one's internal desirability, and it is necessary to find desirable attributes in oneself and express them with 'pride'. Independence is a self-focused model, looking inward and maximizing what is best. Autonomy, personal achievement, and striving for success confirm the internal attributes one perceives, leading to a strong internal motive and active pursuit of acquisition.

Herein the critique we elaborate in the next chapter comes into view: subjective well-being, in its current definition and measurement operationalization, is understood as a positive cognitive and emotional judgment of one's own attributes, state, and environment (Diener, 2000). Moreover, it is assumed to be something acquired by an individual. In other words, the presence of many events that are evaluated as 'good' by one's own self is assumed to be the condition for feeling happiness. Although this approach is widely shared in research emanating out of North America, we can see that the background assumption is an Independent view of human nature, one assuming that a person is defined by his or her internal attributes (including abilities, personality, and experiences gained from education and work experience), and that to become happy, we must maximize and value an individual's internal attributes. But what then happens when people in cultures where the Interdependent mode of happiness dominates, but the surveys they receive ask them to consider their internal attributes, rather than, say, social approval, status, and quality of interpersonal relationships?

# What Explains Such Differences?

To explain such marked differences in how happiness is understood, it is necessary to think conceptually, historically, and philosophically. One of the most important conceptual models to emerge from the field of cultural psychology is the Independent and Interdependent models of self (Markus & Kitayama, 1991). As shown visually in Fig. 4.4, an individual who has an Independent self assumes that the basic unit of the world is an atomized element. Atomized derives from the word atom, a word from physics and chemistry signifying the smallest possible particle that exists. The word 'individual', in turn, comes from the idea that something is not able to be divided, that is, not an amalgamation but a pure substance. This includes the self: one's self has its own boundary and the selves of others have their own boundaries (solid circles in Fig. 4.4a). One's self and those of others are not in a relation of overlap, but of separation and distance. A consequence of the initial assumption that one's self exists independently from others is that the self subsequently works to create relationships with others, in accordance with one's own needs and desires. That is, relationships with others are assumed to be secondary.

In more specialized psychological terms, the independent view of the self is a model in which (1) people are entities with subjectivity defined by attributes (abilities, personality, etc.) that distinguish them from

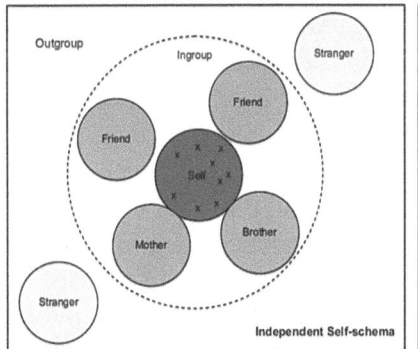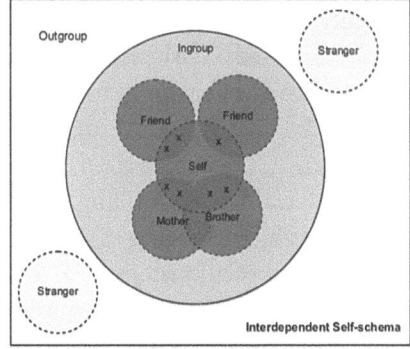

**Fig. 4.4** (a) and (b) Interdependent and independent self (self-construal), schematic conceptualization. (Adapted from Markus & Kitayama, 1991)

others and the surrounding circumstances, (2) the cause of behavior is attributed to these attributes, and (3) interpersonal relationships are subsequently built on the basis of mutual prosocial gain and/or control of the surrounding environment. Here the individual who 'exists well' (well-being) is a subject with high self-esteem, confirmed and expressed through control of his or her surroundings. Therefore, in order to adapt to society, individuals need to confirm that they have desirable attributes and express them with 'pride'. This tendency comes to be further strengthened by the praise they receive from others as a result of expressing their pride. Through this process, values related to 'self-independence' (self-expression, self-assertion, self-esteem, influence over others, etc.) are born. In fact, the tendency to self-aggrandizement (the tendency to think that one is better than others), which is repeatedly verified as a dominant feature of North American personalities, can be thought of as an expression of this tendency. This is empirically verified, for example, in the results of a psychological test called the Twenty Statement Test: US students tend to make many more positive statements about themselves, such as "I am … smart", "I am … a good person", and "I am … good-looking", as compared with other cultural groups.

In contrast, an interdependent self assumes that relationships rather than entities are the constituent elements of the world. One's self and those of others do not have clear boundaries (dotted circles in Fig. 4.4b). There is overlap, co-constitution. That is, one's self and those of others are assumed to co-arise within the relationships among these entities. Entities are ontologically inseparable from webs of relations. More simply: relations are primary, selves secondary. The task for interdependent selves is therefore to pursue collective harmony, to attune to the social matrix within which one is emplaced, as this directly impacts one's 'own' happiness. In more psychological terms, (1) people are part of social relationships that are connected to others and the surrounding circumstances, and thus their definitions of happiness and well-being depend on the nature of the circumstances and interpersonal relationships, (2) their behavior is a result of the circumstances in which they are involved and the reactions of others, and (3) interpersonal relationships are created by the behavior of 'adjusting' to the demands of others. In an interdependent mode, people need to pay attention to their own shortcomings and engage in cooperative relationships with others in order to continually adapt to others and the surrounding situation (Morling et al., 2002). Thus, there is less of a tendency toward highly estimating one's own strength in places like Japan (Heine & Lehman, 1999). In psychological tests we conducted at Kyoto University, in which Japanese participants were asked to write a sentence about themselves in a similar vein to the Twenty Statement Test, the answers emerged as very different compared with the North American samples, that is Japanese students responded 'I am...not sure whether I am smart or not'. For those less familiar with this vast literature, we reproduce one useful table from the seminal 1991 article that further summarizes the key differences (Fig. 4.5).

Markus and Kitayama (1991) suggest that this fundamental difference in self-construal has major consequences for all aspects of psychological functioning: cognition, emotion, and motivation. In terms of emotion, they show a whole range of feelings found in Japanese that are other-focused, and these emotions span the continuum from positive to negative to ambivalent. In our volume, we are extending this paradigmatic

*Summary of Key Differences Between an Independent and an Interdependent Construal of Self*

| Feature compared | Independent | Interdependent |
|---|---|---|
| Definition | Separate from social context | Connected with social context |
| Structure | Bounded, unitary, stable | Flexible, variable |
| Important features | Internal, private (abilities, thoughts, feelings) | External, public (statuses, roles, relationships) |
| Tasks | Be unique<br>Express self<br>Realize internal attributes<br>Promote own goals<br>Be direct: "say what's on your mind" | Belong, fit-in<br>Occupy one's proper place<br>Engage in appropriate action<br>Promote others' goals<br>Be indirect: "read other's mind" |
| Role of others | *Self-evaluation*: others important for social comparison, reflected appraisal | *Self-definition*: relationships with others in specific contexts define the self |
| Basis of self-esteem* | Ability to express self, validate internal attributes | Ability to adjust, restrain self, maintain harmony with social context |

\* Esteeming the self may be primarily a Western phenomenon, and the concept of self-esteem should perhaps be replaced by self-satisfaction, or by a term that reflects the realization that one is fulfilling the culturally mandated task.

**Fig. 4.5** Summary of key differences between an independent and interdependent construal of self. (From Markus & Kitayama, 1991)

conceptual model by looking specifically at how happiness and well-being are influenced by these different forms of self-construal. Our use of the terms Independent Happiness and Interdependent Happiness explicitly acknowledges that our exploration of happiness derives from and furthers the paradigm first laid out by Markus and Kitayama (1991).

Beyond conceptualizing the differences, we may also think historically and philosophically about the roots and routes of these different cultural patterns. What 'caused' such differences? Within the larger field of cultural psychology several persuasive hypotheses have been put forth. Some cultural psychologists focus on the ways different cultures evolved from different bio-regions and forms of food production (e.g., small-scale hunting vs. large-scale rice production). This aligns with a more materialist and evolutionary reading akin to what is found in biology, although shorn of the universalist pretenses of the past. For example, Nisbett and Cohen (1996) suggested that the roots of contemporary cultural differences between North America and the East can be found in the differences between the economic systems of ancient Greece and ancient China. They argue that hunter-gatherer groups came to exalt their own strength and valorized independence, as well as the ability and willingness to forge new paths. This becomes the ideal, according to these scholars, of the American pioneers: there were very few people to rely upon on the frontier. Such circumstances favor a belief in one's own strength as opposed to helping others. It was also an environment where it would be necessary to display one's strength to prevent the theft of cattle or captured prey. These sorts of theories suggest that ecological-material environments of the past created the fundamental patterns we still see playing out today, even despite the vast changes wrought by global integration in intervening centuries.

Similarly, other cultural psychologists suggest that population densities and levels of urbanization are the key factors. They suggest that in the United States, where the population density is relatively low, movement is frequent, and the economic system has historically depended on hunting and gathering, independence tends to be prioritized. Meanwhile in Japan, where the population density is high and the economic system has historically depended on sedentary agriculture, social cooperation tended to be prioritized. Another line of work in this genre is the 'voluntary migration hypothesis', which suggests that individualistic psychological tendencies may be fostered within 'frontier' situations anywhere, given that population densities are low and people have migrated of their own volition, and live without strong cultural support and constraint (Kitayama et al., 2006). Oishi (2010) also focuses on social mobility as a shaper of culture. For most of its history, the United States has enjoyed high social mobility, as evidenced by the number of times people move, change jobs, and high rates of divorce. In such a society, it is more efficient to believe in one's own abilities, to be discerning, and to move in search of newer opportunities. In contrast, in Japanese society, there has been less mobility. Changing jobs or getting divorced was seen as a costly risk, rather than a benefit (e.g., increased opportunities), and thus people tended to be tied to their organizations. In Chap. 6 we return to revisit the question of cultural change, amidst globalization and a changing socio-economic environment.

In contrast to these ideas, other cultural psychologists emphasize cultural factors, particularly millennia-old differences in religio-philosophical approaches to life. Sometimes they use the term 'ideologies' to describe those views, just to ensure that there is no pretense of absolute truth conveyed. It is this latter group who we tend to align ourselves with, which—in turn—brings us into alignment with a range of rich work across the humanities and social sciences. We sketch some of those connections here, as we are keen to show where fruitful collaborative work is possible.

Sometimes, the 'psychology' in cultural psychology creates an image of an empirical approach devoid of historical, social, religious, and philosophical contexts, but in fact *cultural* psychology is—at least in our view—working to empirically validate what humanities and social science scholars have long been pointing out. As discussed previously, *cultural* psychology originally arose from an interdisciplinary dialogue between psychologists and anthropologists. Psychologists brought a scientific, empirical, and quantitative approach to findings of difference surfaced by anthropologists. Undoubtedly, many of the psychological tendencies evident in this empirical data also corroborate sociological studies and resonate with philosophical ideas. Indeed, if the humanities and social sciences remain focused, as they have traditionally done, on examining and understanding the ways humans make sense of their world(s), then the cultural sciences are not in opposition, but complementary. More simply, cultural psychology, by being open to culture, opens the door for cooperation and complementarity, in ways that, say, cognitive psychology premised on universal, a-cultural, essential attributes of the human (singular) psyche has not.

So what religio-philosophical approaches lie behind a notion of independent happiness, one defined by an infinite expansion, maximizing one's internal desirability, and a first assumption of individual independence? The Christian worldview, evolving into Protestantism stands out. It was Christianity that created the notion of individual will—the ability to overcome instinct and nature—and worked to reconstruct the self around the notion of individualized inner distance from one's surroundings, in order to converse with the Absolute (God) (Siedentop, 2014). This was a simultaneous 'disembedding' from one's social surroundings and a re-tethering to a divine realm (Taylor, 2007). In the subsequent move to Protestantism, these latent individual tendencies were intensified. As famously outlined by Max Weber (1905), Calvinism suggested a doctrine of predestination, in which an individual was born one of the 'elect' or 'damned'. God had already decided one's fate, and there was nothing one could do to change it. It was this belief that generated a strong desire for affirming the self as worthy, competent, and true to the intent of God. Herein the desire for acquisition was infinite, as one never really attained ultimate confirmation: it was the drive for affirmation that served as a buffer for the anxiety against the negative prospect of being 'damned'. This manifested in the infinite acquisition of wealth but not in the sharing, consuming, or redistribution of it (hence giving rise to surplus capital that drove the rise of modern capitalism, according to Weber). This 'inner-worldly asceticism' viewed everyday life—one's daily work and conduct—if carried out in the right spirit, as leading to, or at least confirming, one's other-worldly salvation. Individuals would perpetually seek to maximize the 'goodness' of their performance, abilities, and possessions, and feel happy in their prospects for the future.

It was Protestants who attempted to work out their own salvation, a change from the more communal forms of worship dominant in Catholicism. The intensification of individualization—growing conviction around an *entity* view of the person—had been less prominent in

Catholicism. Indeed, even today, the Catholic regions of southern and Eastern Europe tend to show lower levels of individualism than the Protestant dominant areas of Northern Europe, and, in particular, the United States. As for the United States, we must remember that it evolved largely atop the foundation laid by the most radical of Calvinist sects— the Puritans—that were deemed too extreme even for England. It was their worldview that was institutionalized at the outset of the American experiment. In education, for example, both Harvard (1636) and Yale (1701) were founded as training colleges for Puritan church leaders prior to the founding of the country. In fact, nearly all of the leading universities in the United States evolved from strong Protestant roots, and long championed a missionary-style approach to the world (Mardsen, 1994). The first law pertaining to public education was the Massachusetts Old Deluder Satan Act (1636), which argued the need for public schooling to prevent children from falling into the Devil's hands. We are apt to forget in our increasingly secular age, just how much the religio-philosophical scripts of the past still influence the contemporary world. Taylor's (1989) classic work largely concurs, tracing this distinctly Protestant identity all the way back to the inwardness of Greek rationality, showing how the 'radical inwardness' of the Western Enlightenment resulted in 'radical disengagement' from the world, as extended by Descartes and Locke.

In contemporary Western philosophy, there has been growing recognition of the cultural assumptions at work in our ways of seeing the world. For example, in the North American context, John Dewey (1930) wrote the following already a century ago:

> Our moral culture, along with our ideology, is … still saturated with the ideals and values of an individualism derived from the prescientific, pre-technological age. Its spiritual roots are found in medieval religion, which asserted the ultimate nature of the individual soul and centered the drama of life about destiny of that soul. … This moral and philosophical individualism anteceded the rise of modern industry and the era of the machine. It was the context in which the latter operated. … But the fact that the controlling institution was the Church should remind us that in ultimate intent it existed to secure the salvation of the individual. (Dewey, 1930)

Dewey's philosophical and educational project consisted in trying to move beyond such atomized individualism, and place a greater emphasis on collective approaches. Unfortunately, Dewey's pragmatism, which actually carries strong resonances with the Interdependent mode, lost out in the United States to further individualistic, technocratic approaches (Labaree, 2010). We come back to discuss this further in the next chapter.

In contemporary European philosophy, we find deeper recognition of the enduring impact of Protestantism. Following in a line of critique first laid by Nietzsche, Gadamer (2006 [2000]) concluded his illustrious career with a call to understand how different cultural spheres, emerging from other religious worldviews, comprehend the modern liberal West:

> Let us ask, rather, "How is the Enlightenment comprehended among these religions?" Even now, you see, even now it is clear to me that the same thing by no means follows from the worldview of Japan, for example, as it does for us. It's really like this—in the great chain of the experience of transcendent, only one experience has been salvaged in our [Western] case, and that, to be precise is Calvinism. (p. 73)

Although the philosophical language here is far removed from the heavily empirical work of cultural psychology, the underlying message is largely congruent: that the individualism of the West, particularly pronounced in the United States, evolved out of the religio-philosophical conditions of Western Christendom, particularly its later Protestant radicalization.

So then where does the interdependent view found across East Asia come from? The East Asian cultural composition comprises a mix of Confucian, Daoist, and Buddhist influences. As seen in the Vinegar Tasters, each of these different traditions has played a constitutive role in the East Asian worldview, and each emphasizes a different view of happiness. Yet, we would argue that, at root, the teachings of these three traditions are surprisingly congruent, suggesting different emphases along a broadly similar worldview, rather than vastly divergent approaches. As one would expect, these 'teachings' are complementary, rather than in conflict. This helps explain the constant co-mingling and cross-fertilization between the three traditions over the past 1500 years. This root similarity is found in the lack of emphasis on maximizing one's own self and personal desires. Confucianism understands identity only within a matrix of relationships, Daoism understands self as a manifestation of a larger cosmic flux, and Buddhism—perhaps the most radical in its discussion of self—seeks to remove the illusion of self all together. In all three approaches, the focus is 're-embedding' self within a given context—social, ecological, metaphysical, and so on. Put differently, all three traditions suggest the impossibility of starting with a de-contextualized view of self. Each approach teaches interdependence with the surrounding environment. From within such a worldview, the notion of 'internal desirability' can only be answered *in relation to* the context in question. Moreover, in trying to acquire something that is desirable, consideration must be taken into account of how the acquisition might impact or effect other elements *in the relationship*. Instead of gratification of entities, we have balancing of trade-offs *within relationships*. The role of the Other (i.e., what is beyond the self) cannot be overestimated here. In Confucianism, a distinction is made between the small-self and the big-self, the former signifying egoism. In some forms of Buddhism, such as the Pure Land Schools in Japan, the Other is literally viewed as means to salvation. Across much of East Asia then, which never experienced the inward turn of the Platonic-Protestant Western tradition, the focus remains resolutely Other-centered (allo-centric).

Pushing a little deeper, both Daoism and Buddhism explicitly emphasize a world in constant flux. Impermanence and change are key motifs. This stands in contrast to the Platonic-Christian tradition of permanence (true Forms; eternal Heaven; everlasting Peace). Moreover, there is dominant view in East Asia that identity—of anything, not just self-identity—only arises in relation. For example, beauty only exists in relation to that which is ugly; black only has meaning in relation to white. When these two dimensions are put together, there emerges a belief that a given state cannot persist indefinitely, but will inevitably reverse into its opposite. Within such a worldview, it makes more sense to remain in the middle, constantly attentive to change, awaiting a shift that will inevitably come. Here the view of an unalloyed 'good' that continues infinitely makes no sense. We will review below empirical evidence that strongly suggests that such views continue to underlie a Japanese construal of what happiness 'is'.

In terms of Japan, many consider the Zen Buddhist schools that were culturally dominant for centuries as a hybrid of Buddhism and Daoism. Buddhist ideas, borrowed from India, were understood atop a Daoist worldview of dynamic opposites creating the whole (Deguchi et al., 2021). In the twentieth century, when Japanese culture came into contact with the West, the (re)articulation of this worldview arose, as the Zen-Daoist worldview did not sit well with Western modernity's WEIRD first assumptions. In the context of the current volume, Japanese (re)articulations of psychology at this time are particularly insightful.

Nishida Kitaro (1870–1945) who is widely considered Japan's most influential philosopher drew heavily on Zen to formulate a creative response to Western thought. Among his first works were a critical examination of William Wundt's (1896) *Grundriss der Psychologie* (Outline of Psychology), a work that laid out the assumptions, methods, and concerns of the contemporary science of psychology the Western world still practices today. Nishida, in his initial 1904 lectures, took issue with Western psychology's first assumption of an individual who cognizes, feels, and wills autonomously. Nishida's critique, formulated with inspiration from Zen-Daoist ideas, ultimately focused on the embeddedness of the individual. "Although I am myself, I do not determine myself alone. I am determined thoroughly by the Other", Nishida constantly stressed, continuing "We are not determined only by the inside, as the psychologists try to convince us. The self cannot be determined without relations to the outside" (Nishida, 1932). Nishida even argued that "the inside alone is a mere fantasy, and a fantasy is not the true self. It is within the connection between the outside and inside where the true meaning of self lies" (ibid.). Perhaps no better religio-philosophical articulation of the interdependent mode can be found than this statement.

To shift attention away from the independent self, Nishida appealed for attention to the 'place' (basho) which gave rise to entities. Many scholars have pointed out that this 'place' signifies a place of fundamental relation (Sevilla, 2017; Kasulis, 1998). Nishida's ideas around an encompassing self and *basho* spawned a whole tradition of philosophy that went on to impact the Japanese social sciences. Many of these ideas, later taken up by Japanese psychologists such as Takeo Doi and Hayao Kawai, actually produced some of the ideas that helped advance cultural psychology—a distinctively Japanese response to the Western worldview embedded in mainstream psychology. The volume you hold in your hands right now is evidence of just how impactful non-Western religio-philosophical approaches remain in some parts of East Asia.

Before coming back to happiness and well-being, it is worth addressing here that there is an inherent tension involved in elaborating the roots of these differences. On the one hand, we do not view the interdependent model of happiness as unique to East Asia. Yet, on the other hand, we recognize its roots in the religio-philosophical tradition of the East Asian region. Is there not a contradiction? Here again we are emphasizing dominant patterns. We believe that even in the North American context, cultural ideas approximating interdependent modes can be found, such as in Dewey's pragmatism, feminist theory, and indigenous worldviews (the 'Americans' there before Protestant Europeans arrived). Indeed, Nishida felt resonances with the American pragmatist William James, and the similarities between, say, eco-feminism and Daoism are, in many respects, rather stunning (Silova, 2020). Yet, these different cultural frameworks have never become dominant there, probably as a result of the Platonic-Christian (Protestant) complex *that is the 'unique' story of the Western, particularly Anglo-American, world* (Rappleye, 2018). By extension, religio-philosophical systems such as Ubuntu in Africa show striking resemblances, but are only now emerging from Western colonization. Our view is not that East Asia is unique, but that the Interdependent themes that are clearly articulated in its religio-philosophical traditions were institutionalized across the region for millennia (e.g., Confucian academies in China, *terakoya* Buddhist temple schools in Japan), and—crucially—these traditions have been less impacted by the Platonic-Christian worldview, even under modernity (Rappleye, 2024). From this angle, the interdependence approach we are working to lay out here helps re-actualize traditions elsewhere around the world that have been forgotten or sidelined due to the dominance—still ongoing today, as we shall see in the next two chapters—of the independent mode.

# Back to the Differences: Qualities of Happiness and Well-Being

Let us now return to concrete differences in happiness, sharing empirical work that confirms the conceptual and theoretical discussion thus far. In terms of internal attributes versus relations, as we reviewed above, Diener and Diener (1995) conducted a comparative study in 31 countries using the individualism-collectivism axis and found that the impact of self-esteem on subjective well-being is stronger in cultures that emphasize the 'individual' such as the West than in cultures that emphasize the 'group'. In Japan relationships are the crucial element in happiness, and moreover the availability of emotional support from close people has been found to be particularly related to levels of happiness (Uchida et al., 2008; Uchida & Kitayama, 2009). In addition, pleasant emotions (such as familiarity) that are obtained when in harmonious relationships with others are more associated with happiness in the Japanese context. Relationships within local networks and relationships within the workplace are also major factors in Japanese happiness. Emotional support refers to receiving love and affection from those around us, and receiving support in various aspects of our lives when we are in need. Interestingly, in North America, which places a high value on self-esteem, receiving support can actually lead to an awareness of one's own powerlessness, which can threaten one's self-esteem and lessen one's sense of well-being. In Japan, by contrast, the relationship between happiness and support has been shown to be stronger because support is a recognition of interpersonal bonds, even if receiving support damages one's evaluation of being in control (self-efficacy, self-esteem) (Uchida et al., 2008).

In terms of meaning, in a comparative study of Japan and the United States conducted by Uchida and Kitayama (2009) detected fundamental differences and valences assigned to the key terms. The results are shown in Fig. 4.6. When asked to describe the meaning of happiness in five ways, 97.4% of the total descriptions obtained in the United States were positive (e.g., when you achieve something, you feel like jumping up and down, you feel positive about everything, you can be kind to others, and your self-esteem increases). In contrast, in Japan, only 68% of the statements were positive, with the remaining 30% suggesting negative

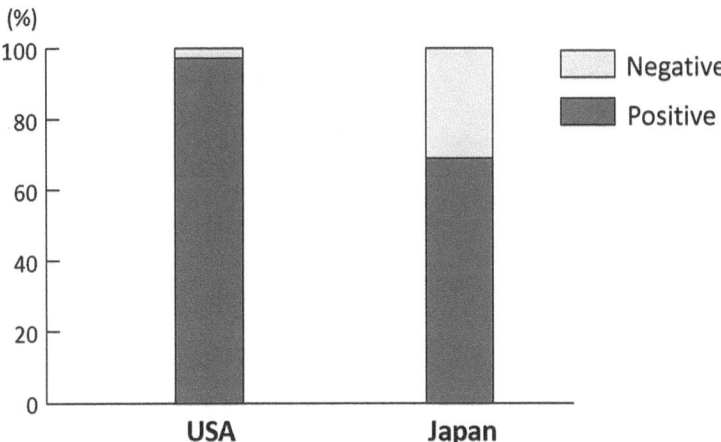

Fig. 4.6 Evaluation of the meaning of happiness. (Adapted from Uchida & Kitayama, 2009)

dimensions: "If I am happy, people will be jealous of me", "I won't be able to care for others", "If I am too happy, I won't be able to grow", and "I will eventually lose it". Conversely, a similar survey on unhappiness revealed that 90% of the American respondents described unhappiness in a negative light, while about 30% of the Japanese respondents found positive elements, such as "unhappiness has beauty" and "unhappiness can be an opportunity for self-improvement". As we will discuss in the next chapter, if the ideal level of happiness in East Asia falls within this more moderated range, then moderate scores on subjective well-being may not be 'low' but instead manifest culturally mediated reflections about the ideal level of happiness.

In the same study, the meaning of happiness was mapped in Japan. Respondents were asked to freely associate happiness with any other word or phrase that came to mind. As shown in Fig. 4.7, we may think of these responses along two axes: ego-centric versus allo-centric, and outwardly expressed versus inwardly felt. As we can see, some of the familiar forms of independent happiness are present in the Japanese responses, such as in the equating of money with happiness and/or material possessions. But what is most striking here are the other meanings assigned by respondents. Many equated happiness with relations and allo/other-centric

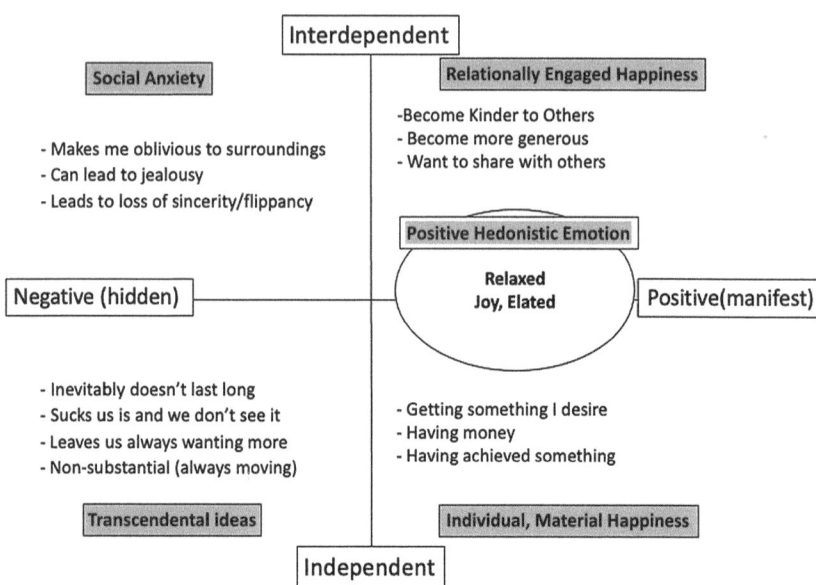

**Fig. 4.7** Map of meaning of happiness in Japan. (Adapted from Uchida & Kitayama, 2009)

feelings such as generosity or sharing. In a more negative valence, happiness was associated with anxiety, the idea that one's happiness might spark jealousy and/or undue attention. But whether positive or negative, happiness was understood *in relation*. We also see here a sense of happiness as something impermanent: something that would not last long or was inevitably fleeting.

It is worth looking at impermanence and balance in more depth, as it reveals much and brings together the wide ranging discussion of the current chapter. In North America, happiness is defined as the state in which one's abilities and possessions (physical and social) are maximized. At work here is an assumption that further positive traits and favorable circumstances will allow one to attain even higher states of happiness. Yet, the sorts of assumption leading to the North American infinite, acquisitive model of happiness are not as operative in Japan. As we have seen, happiness carries both positive and negative aspects. Emplaced in a fundamental worldview focused on change, the logical conclusion is that it is difficult to be happy all the time, and too much happiness would, in any case, invite the negative aspects of happiness to emerge. Good and bad occur side-by-side; life oscillates between and encompasses the two poles. Within such a view, a focus on balance comes to the fore: avoiding extremes of too much or too little happiness; accepting that good and bad come together or follow in sequence. This balance orientation approximates Daoist principles of one extreme leading to the next, and co-arising—a yin-yang view of happiness. Yin and Yang means, in the original Chinese, dark or hidden (yin) and bright or revealed (yang). What is hidden will, as a result of change, come to light, and that which is revealed will, with time, recede. Despite modernization of the outward institutions of East Asian societies, these sorts of fundamental worldviews are still at play, as this 'traditional' wisdom has proven to be surprisingly durable for coping with modern life.

Interestingly, this Yin-Yang approach is not merely fancy conceptualization, but derives from empirically validated differences in the ways East Asians and North Americans predict change. In one well-known experiment, line graphs showing time series changes in several patterns were shown to American and Chinese study participants, as shown in Fig. 4.8. Participants were then asked to predict the subsequent changes. When doing so, Americans were more likely to predict in line with the previous changes (if the graph showed an increasing trend, they would predict an increase), while the Chinese were more likely to predict the opposite of the previous changes (if the graph showed an increasing trend, they would predict a decrease). Furthermore, when several graphs of linear and non-linear changes were shown and people were asked to judge which they thought was the happiest life if of that line were to represent life, linear changes were considered better by Americans, while non-linear changes were considered better by Chinese (Ji et al., 2001).

These sorts of beliefs and values also deeply influence emotional experiences. When measuring the intensity of positive and negative emotions, research has also shown that in the United States, they are negatively correlated, while in East Asian cultures (China and Korea), they are positively correlated (Bagozzi et al., 1999). In other words, while positive and

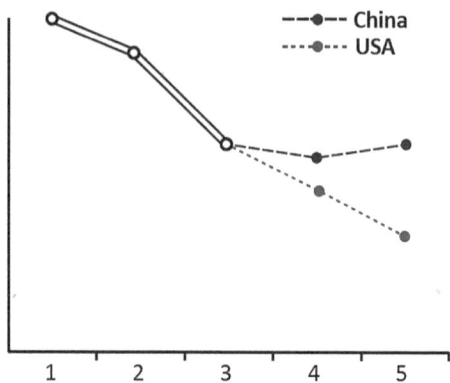

**Fig. 4.8** Prediction of change for Americans and Chinese (the first three points of change were presented first, and then participants were asked to predict the next two points. The Americans predicted that the first three points of change [downward] would continue, while the Chinese tended not to predict the same linear change. (Modified from Ji et al., 2001))

negative emotions are polarized in the West, they coexist in a balanced manner in East Asian cultures. In fact, it has been reported that in Japan, there is a high occurrence of ambiguous feelings of something being both "positive and negative" (Miyamoto et al., 2010), a result we may read as the epitome of balance.

# References

Adams, G., & Markus, H. R. (2004). Toward a Conception of Culture Suitable for a Social Psychology of Culture. In M. Schaller & C. S. Crandall (Eds.), *The Psychological Foundations of Culture* (pp. 335–360). Erlbaum.

Arnett, J. (2008). The Neglected 95%; Why American Psychology Needs to Become Less American. *American Psychologist, 63*, 602–614.

Bagozzi, R. P., Wong, N., & Yi, Y. (1999). The Role of Culture and Gender in the Relationship Between Positive and Negative Affect. *Culture and Emotion, 13*, 664–672.

Brumann, C. (1999). Writing for Culture: Why a Successful Concept Should Not Be Discarded. *Current Anthropology, 40*, S1–S27.

Deguchi, Y., Garfield, J., Priest, G., & Sharf, R. (2021). *What Can't Be Said: Paradox and Contradiction in East Asian Thought*. Oxford University Press.

Dewey, J. (1930, February 19). *Individualism: Old and New*. New Republic. https://vdoc.pub/documents/individualism-old-and-new-1mss4u3fpqbo

Diener, E. (2000). Subjective Well-Being: The Science of Happiness and a Proposal for a National Index. *American Psychologist, 55*, 34–43.

Diener, R., & Diener, M. (1995). Cross-cultural Correlates of Life Satisfaction and Self-Esteem. *Journal of Personality and Social Psychology, 68*, 653–663.

Diener, E., & Suh, E. M. (2000). *Culture and Subjective Well-Being*. MIT Press.

Endo, Y. (1995). Argument of the Self as an Index of Mental Health. *Japanese Journal of Social Psychology, 11*, 134–144.

Gadamer, H-G. (2006 [2000]). *A Century of Philosophy: A Conversation with Riccardo Dottori*. Continuum (Originally Published in Germany '2000) by LIT Verla, Hamburg).

Heine, S. J., & Lehman, D. R. (1999). Culture, Self-Discrepancies, and Self-Satisfaction. *Personality and Social Psychology Bulletin, 25*(8), 915–925.

Heine, S. J., Kitayama, S., Lehman, D. R., Takata, T., Ide, E., Leung, C., & Matsumoto, H. (2001). Divergent Consequences of Success and Failure in Japan and North America: An Investigation of Self-Improving Motivations and Malleable Selves. *Journal of Personality and Social Psychology, 81*(4), 599–615.

Hendriks, T., Warren, M. A., Schotanus-Dijkstra, M., Hassankhan, A., Graafsma, T., Bohlmeijer, E., & de Jong, J. (2019). How WEIRD are Positive Psychology Interventions? A Bibliometric Analysis of Randomized Controlled Trials on the Science of Well-Being. *The Journal of Positive Psychology, 14*(4), 489–501.

Henrich, J., Heine, S. J., & Norenzayan, A. (2010). The Weirdest People in the World? *Behavioral and Brain Sciences, 33*(2–3), 61–83.

Iyengar, S. S., & Lepper, M. R. (1999). Rethinking the Value of Choice: A Cultural Perspective on Intrinsic Motivation. *Journal of Personality and Social Psychology, 76*(3), 349–366.

Ji, L., Nisbett, R. E., & Su, Y. (2001). Culture, Change, and Prediction. *Psychological Science, 12*, 450–456.

Kasulis, T. (1998). *Intimacy or Integrity: Philosophy and Cultural Difference*. University of Hawaii Press.

Kitayama, S. (1998). Jiko to kanjyo: bunka shinrigaku ni yoru toikake [*Culture and Emotion: Questions from Cultural Psychology*]. Kyoritsu Shuppan.

Kitayama, S., & Markus, H. (2000). The Pursuit of Happiness and the Realization of Sympathy: Cultural Patterns of Self, Social Relations, and Well-Being. In E. Diener & E. M. Suh (Eds.), *Culture and Subjective Well-Being* (pp. 113–161). MIT Press.

Kitayama, S., Takagi, H., & Matsumoto, H. (1995). Casual Attribution of Success and Failure: Cultural Psychology of the Japanese self. *Shinrigaku Hyoro [Japanese Psychological Review], 38*(2), 247–280.

Kitayama, S., Markus, H. R., Matsumoto, H., & Norasakkunkit, V. (1997). Individual and Collective Processes in the Construction of the Self: Self-Enhancement in the United States and Self-Criticism in Japan. *Journal of Personality and Social Psychology, 72*(6), 1245–1267.

Kitayama, S., Ishii, K., Imada, T., Takemura, K., & Ramaswamy, J. (2006). Voluntary Settlement and the Spirit of Independence: Evidence from Japan's 'Northern Frontier'. *Journal of Personality and Social Psychology, 91*(3), 369–384.

Kroeber, A. L., & Kluckhohn, C. (1952/1966). *Culture: A Critical Review of Concepts and Definitions.* Vintage.

Labaree, D. (2010). How Dewey Lost: The Victory of David Snedden and Social Efficiency in the Reform of American Education. In D. Thohler, T. Schlag, & F. Osterwalder (Eds.), *Pragmatism and Modernities* (pp. 163–188). Sense Publishers. http://www.web.stanford.edu/~dlabaree/publications/How_Dewey_Lost.pdf

Mardsen, G. (1994). *The Soul of the American University: From Protestant Establishment to Established Nonbelief.* Replica Books.

Markus, H. R., & Kitayama, S. (1991). Culture and the Self: Implications for Cognition, Emotion, and Motivation. *Psychological Review, 98*, 224–253.

Miyamoto, Y., Uchida, Y., & Ellsworth, P. C. (2010). Culture and Mixed Emotions: Co-occurrence of Positive and Negative Emotions in Japan and the United States. *Emotion, 10*, 404–415.

Morling, B., Kitayama, S., & Miyamoto, Y. (2002). Cultural Practices Emphasize Influence in the United States and Adjustment in Japan. *Personality and Social Psychology Bulletin, 28*(3), 311–323.

Myers, D. G., & Diener, E. (1995). Who Is Happy? *Psychological Science, 6*, 10–19.

Nisbett, R., & Cohen, D. (1996). *Culture of Honor: The Psychology of Violence in the South.* Westview Press.

Nishida, K. (1932). Jitsuzon no konteitoshite jinkaku gainen [The Concept of Personality as the Foundation of Reality (Lecture Delivered at the Shinano Education Center from 3–5 September)]. In N. Kitaro (2020) *Collection of Lectures by Nishida Kitaro.* Iwanami Press (pp. 19–32).

Oishi, S. (2002). The Experiencing and Remembering of Well-Being: A Cross-Cultural Analysis. *Personality and Social Psychology Bulletin, 28*(10), 1398–1406.

Oishi, S. (2010). The Psychology of Residential Mobility: Implications for the Self, Social Relationships, and Well-Being. *Perspectives on Psychological Science, 5*(1), 5–21.

Oishi, S., & Diener, E. (2001). Goals, Culture, and Subjective Well-Being. *Personality and Social Psychology Bulletin, 27*, 1674–1682.

Rappleye, J. (2018). Borrowings, Modernity, and De-Axialization: Rethinking the Educational Research Agenda for a Global Age. In A. Yonezawa, Y. Kitamura, B. Yamamoto, & T. Tokunaga (Eds.), *Japanese Education in a Global Age: Sociological Reflections and Future Directions* (pp. 53–74). Springer.

Rappleye, J. (2020). Comparative Education as Cultural Critique. *Comparative Education, 56*(1), 39–56.

Rappleye, J. (2024). Modernity and Education. In M. Ueno, M. Okabe, & F. Ono (Eds.), *Philosophy of Education in Dialogue Between East and West: Japanese Insights and Perspectives* (pp. 139–160). Routledge.

Rappleye, J., & Komatsu, H. (2017). How to Make Lesson Study Work in America and Worldwide: A Japanese Perspective on the Onto-cultural Basis of (Teacher) Education. *Research in Comparative and International Education, 12*(4), 398–430.

Sevilla, A. (2017). *Watsuji Tetsuro's Global Ethics of Nothingness: A Contemporary Look at A Modern Japanese Philosopher.* Palgrave.

Siedentop, L. (2014). *Inventing the Individual: The Origins of Western Liberalism.* Harvard University Press.

Silova, I. (2020). Anticipating Other Worlds, Animating Our Selves: An Invitation to Comparative Education. *ECNU Review of Education, 3*(1), 138–159.

Suh, E. M. (2002). Culture, Identity Consistency, and Subjective Well-Being. *Journal of Personality and Social Psychology, 83*, 1378–1391.

Suh, E. M., Diener, E., Oishi, S., & Triandis, H. C. (1998). The Shifting Basis of Life Satisfaction Judgments across Cultures: Emotions versus Norms. *Journal of Personality and Social Psychology, 74*, 482–493.

Taylor, C. (1989). *Sources of Self.* Harvard University Press.

Taylor, C. (2007). *A Secular Age.* Harvard University Press (Belknap).

Triandis, H. C. (1995). *Individualism and Collectivism.* Westview Press.

Uchida, Y., & Kitayama, S. (2009). Happiness and Unhappiness in East and West: Themes and Variations. *Emotion, 9*, 441–456.

Uchida, Y., & Ogihara, Y. (2012). Bunkateki kofukukan: bunkashinrigakuteki chimi to shourai he no tennbo [A Cultural View of Happiness: Findings and Futures from a Cultural Psychology Approach]. Shinrigaku hyoron [*Japanese Psychological Review*], *55*, 26–24.

Uchida, Y., Norasakkunkit, V., & Kitayama, S. (2004). Cultural Constructions of Happiness: Theory and Empirical Evidence. *Journal of Happiness Studies, 5*, 223–239.

Uchida, Y., Kitayama, S., Mesquita, B., Reyes, J., & Morling, B. (2008). Is Perceived Emotional Support Beneficial? Well-being and Health in Independence and Interdependent Cultures. *Personality and Social Psychology Bulletin, 34*, 741–754.

Weber, M. (1905). *The Protestant Ethic and the Spirit of Capitalism* (T. Parsons, Trans.). Allen & Urwin. (1930 English Translation)

# 5

# An Interdependent Approach: Manifestations in Cultural Practices

In the previous chapter, we elaborated on the interdependent mode of happiness. We sought to provide conceptual constructs, theoretical exploration, and empirical support. In this chapter, we extend that discussion to the larger puzzle of culture. In doing so, we return to questions raised in Chap. 2: the relationship between 'micro' self-construal and 'macro' contexts. How does an interdependent mode manifest in the larger cultural patterns? What sorts of alternative cultural practices does an interdependent approach give rise to? How do cultural practices help reinforce or, in some cases, challenge dominant patterns? In line with the discussions thus far, we focus on three specific cultural domains: measurement, education, and social capital. In doing so, we link cultural psychology's focus on modes of self-construal with larger patterns of social-construal, traditionally the domain of social sciences disciplines like sociology. This further extends the interdisciplinarity of the last chapter, showing how psychology, philosophy, and social science—or at least the culturally aware communities within these broad academic fields—come to be mutually reinforcing. This combined micro and macro approach is a crucial synthetic perspective necessary for understanding the future of happiness and well-being.

© The Author(s) 2024
Y. Uchida, J. Rappleye, *An Interdependent Approach to Happiness and Well-Being*,
https://doi.org/10.1007/978-3-031-26260-9_5

# Preliminary Considerations: Culture→Mind? Mind→Culture? or Culture⇔Mind?

In psychology, 'psychological function' is the term used to encompass the totality of how the mind 'works'. Psychological functioning includes thinking and feeling, as well as motivation and behavior (action). These psychological functions are, at least to some extent, clearly subject to biological limitations. For example, thinking and feeling are limited by heredity effects on how brain and body function. Moreover, there are undoubtedly many commonalities in psychological functioning among humans worldwide. Nevertheless, the specific environment, circumstances, and personal experiences we undergo also have a significant effect on psychological function. Cultural psychology is one part of a much larger movement across the human sciences that has, in recent decades, shown the severe limitations of biological-heredity style, universalist explanations that dominated the first half of the twentieth century.

Among the specific environments and circumstances that impact psychological function, one element is culture. In the last chapter, we tentatively defined culture as "a pattern of values, thoughts, and reactions that has been developed and shared by a group of people throughout their social and group history". The questions we now address are these: How is this culture transmitted, learned, and—once established—continually reinforced? As we shall see, cultural values and dispositions are transmitted at both the macro-level—for example, the political, economic, educational, religious, and linguistic scripts—prevailing in a given society, and at the meso-level—immediate relationships at school, home, work, and/or places of worship. Once a mutually reinforcing cycle of psychological function and cultural values is in place, people become involved in the maintenance of cultural practices simply by performing habitual acts. More accurately, habitual actions transmit such cultural values, while heretical acts work to transform them.

Within the wider 'macro' social sciences, this discussion has unfolded under the so-called structure versus agency debate. This debate has been particularly strong in Anglophone social science, where the predominant fault lines have been Marxism and Liberalism. Under a structuralist account, the wider structures of a given society limit the choices of the individual. This includes the constraints of social class, race, gender, education, religion, language, and so on. A Marxist account emphasizes the ways that structures impute economic power, and socialize individual 'minds' into ways of thinking that constrain their freedom. Theorists like Gramsci, Lukacs, and Althusser, mostly building on Engels, often utilize the term 'false consciousness' to describe these mistaken, ideological ways of thinking. In this account, 'culture' is seen as the key element of the ideological apparatus of control, and embraces the assumption that 'culture' shapes 'mind' unilaterally. Culture is a structure, imposed and hegemonic, that controls unwitting individuals.

In contrast, an agency account suggests that individuals have considerable autonomy, are far less constrained by wider social structures than imagined, and through their own abilities, intelligence, and action can change these wider social structures. This is, roughly speaking, the dominant view of Liberalism. By exercising human reason, the 'minds' of individuals can shape their larger environment, both social institutions and/ or natural environments. Here the view is that 'mind' wholly shapes culture: the faculty of Reason is viewed, in these liberal accounts, as universal and the source of independence. When exercised properly, reason is not influenced by the prejudices of culture. Out of this a-cultural use of reason springs, of course, the universal claims of the Western Enlightenment, a social and political movement premised on the universality of reason and the individual agency it affords. In this way, Anglophone social sciences present us with a dichotomy: either culture shapes mind or mind shapes culture. We note, in passing, that recent debates in Continental Europe are somewhat richer and more nuanced, following poststructuralist challenges to a-cultural reason (Foucault among others) and greater reflexivity on the cultural contingency of such categories.

Cultural psychology takes a divergent approach, one emphasizing the "mutual constitution" of culture and self (Shweder, 1991). Markus and Kitayama (2010) here provides another useful visual conceptualization, as shown in Fig. 5.1. Here we may envisage how the 'self' is influenced by factors at various levels: from daily practices at the meso-level, to institutions and pervasive ideas at the macro-level. The notion of self here includes cognition, emotion, and motivation, but also perception (what

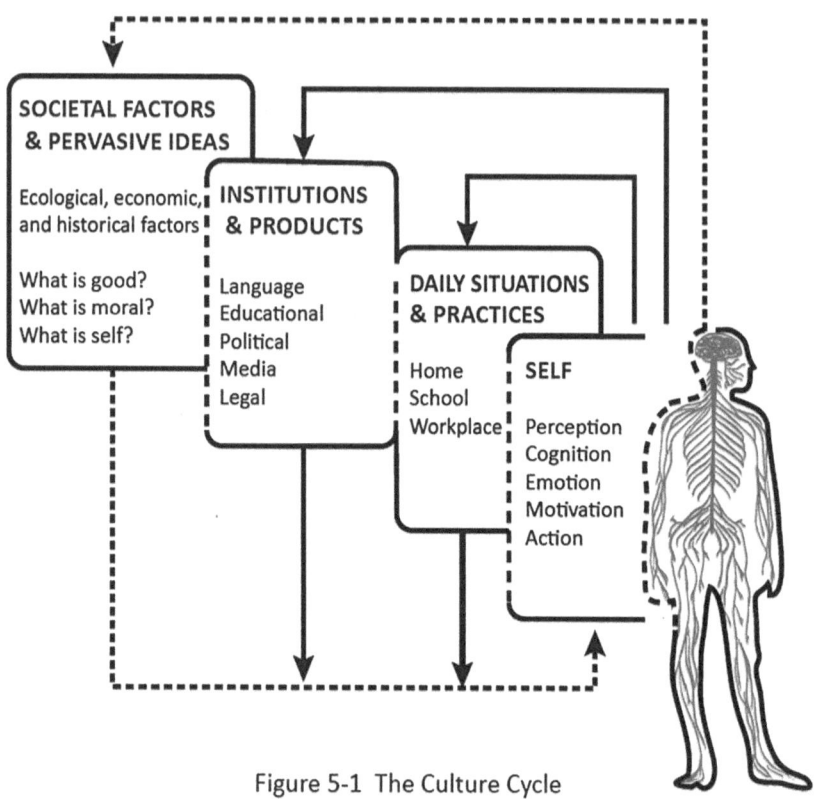

Figure 5-1 The Culture Cycle

**Fig. 5.1** Mutual constitution of culture and selves. (Adapted from Markus & Kitayama, 2010)

becomes our focus) and action (what moves us). This self is formed by and forms the daily practices we find ourselves in. That is, the properties of 'mind' are created through participation in "everyday habits and realities" in familiar relationships, such as family, school, and workplace, where various concrete tasks are carried out and learned through repeated actions (Kitayama et al., 2009). In turn, these tasks constitute the larger linguistic, political, educational, legal, and media 'scapes' we find ourselves embedded within; the totality of customary and public semantic structures, folk theories, and symbols built up through the history of a society or group. And from these scapes come "cultural products"

(Morling & Lamoreaux, 2008), which would include news articles, song lyrics, textbooks, children's books and so on. All of these come back around to reinforce those dispositions in our selves.

At the most pervasive yet intangible level, larger factors such as economic patterns and environmental landscapes reinforce these patterns. A neo-liberal capitalist economy that exhorts individuals to increase the career prospects by *acquiring* higher skills (human capital stocks) and strategically building a *unique* skillset (e.g., the resume) produces a vastly different outlook than a gift-economy founded on reciprocal exchange as means of affirming solidarity (Mauss, 1898 [2000]). Cramped single family apartments in densely populated yet anonymous urban sprawls, as opposed to multigenerational homes in one's ancestral home surrounded by intimate relations and nature, inevitably gives rise to very different emotional and perceptual experiences of the self.

Here too we may locate philosophical ideas, religious doctrines, and folk theories that are often shared, but rarely articulated. In the last chapter, we saw that modern Japanese philosophy emphasized that "the self cannot be determined without relations to the outside" (Nishida, 1932). Contrast that with Dewey's observations that American individualism finds its 'spiritual roots' in Christianity. Philosophical and religious ideas such as these both shape and are shaped by the wider cultural milieu. Again, once the psychological processes are established, people consciously or unconsciously participate in the maintenance of acquired cultural practices through their participation in familiar collective phenomena, even something as mundane as non-verbal communication (e.g., a bow vs. a handshake) and/or acting together on a common task. Each loop or layer of the system works to keep the other in place, creating considerable inertia. That said, transformation can also occur, particularly when differences enter the milieu and are sustained over longer periods of time.

In this way, cultural psychology conceptualizes one's mind (self) as not completely independent of social and cultural customs/products. Social and cultural customs/products do not, in turn, exist apart from one's mind and actions. While this approach is at odds with dominant models in contemporary Anglophone social science, it is interesting to note that leading Western thinkers like John Dewey held a strikingly similar view. Returning to Dewey's insights shared in the last chapter, we find Dewey underscoring the way that the 'deep-seated individualism' of American society has indelibly shaped Anglo-American institutions:

> The early phase of the industrial revolution wrought a great transformation. It gave a secular and worldly turn to the career of the individual, and it liquified the static property concepts of feudalism by the shift from agriculture to manufacturing. Still, the idea persisted that the property and reward were intrinsically individual …. a fusion of individual capitalism, of natural rights, and of morals founded on strictly individual traits and values remained, under the influence of Protestantism, the dominant intellectual synthesis. (Dewey, 1930)

Here religious individualism gave rise to these secular institutions, but once in place these institutions come to reinforce an individualistic, independent view of self. In the earlier discussion of Anglophone social science, theories of liberal agency and autonomy are themselves part of the 'cultural products' that function to reinforce such mental models. That is, our academic theorizing itself is not distinct from this culture cycle, as Dewey well recognized.

It may be useful to illustrate further with a brief empirical example. In one study, Markus, Uchida and others (Markus et al., 2006) compared how Olympic athletes' motivations and emotions are covered in the media coverage in Japan and the United States. In Study 1, they examined the word-for-word content of Japanese and U.S. newspaper, television, and magazine coverage (including athlete commentary, reporter analysis, and commentator analysis) of 77 Japanese and 265 American athletes at the 2000 Sydney Olympics and the 2002 Salt Lake City Winter Olympics. Figure 5.2 shows the results of this analysis.

Here American news articles tended to mention personal characteristics such as 'athletes' abilities and personalities' and 'competitiveness and rivalry' with greater frequency than the Japanese articles. In contrast, Japanese news articles tended to mention 'others' (family, coaches, friends) more often than American articles, and to focus on bringing joy to the people around them and their fans as a major motivator. In addition, news content that focused on the 'athletes' emotions', such as their individual moods in competition situations, was also mentioned more

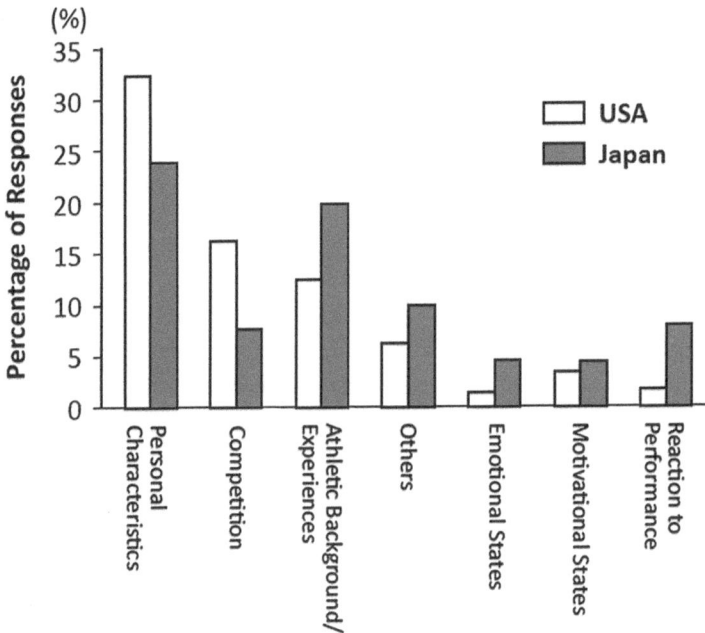

**Fig. 5.2** Media coverage of Olympic athletes in Japan and the USA. (Adapted from Markus et al., 2006)

often in the US. The study also looked at the amount of coverage of positive factors such as 'talented' and 'former champion' versus negative factors such as 'weak-minded' and 'had a hard time'. In the United States, the media paid little attention to negative elements, whereas in Japan, the media covered the negative elements in roughly equal proportion to the positive elements.

Stepping back to view this conceptually, reporters writing these stories were influenced by culture, and those reading the stories had their cultural dispositions about what leads to 'good' performance further reinforced. We can imagine readers would go on to attribute their own success in more mundane tasks in similar ways—individual strength for Americans, an affordance by others among Japanese. That is, the cultural product of the Olympic news coverage produced and was a product of the larger dispositions in these two divergent cultural contexts. That is, mind and culture co-constitute.

# (Mis) Measurement Revisited: Happiness Rankings as Cultural Product

Chapter 2 devoted considerable time to outlining the different means of measuring and ranking happiness emerging over the past decade, those following in the wake of a collective loss of faith in the GDP=Happiness equation of the twentieth century. We showed that on some rankings such as the SPI, Japan scored rather high. However, on other rankings such as the WHR, Japan and Korea scored poorly, at least in comparison with other high-income countries. These differences in 'ranks' largely come down to whether or not subjective well-being is included, and how much it is weighted in the total score: the more a given rank leans on subjective well-being, the lower Japan and East Asia tend to score. Indeed, on the Cantril Ladder (*Please imagine a ladder with steps numbered from zero at the bottom to 10 at the top…*) the average score for Japan is consistently around 6.5. These results rarely fluctuate, no matter whether it is an international survey or large-scale domestic surveys. Figure 5.3 illustrates this, showing scores over three consecutive years (2009–2011), for a representative sample of 4000 Japanese citizens aged 15–80 on a ten-point Cantril-style life satisfaction scale (Cabinet Office, Government of Japan, 2011). These years were a highly eventful time that included the Global Financial Crisis of 2008 and Japan's Triple Disaster of 2011. Yet

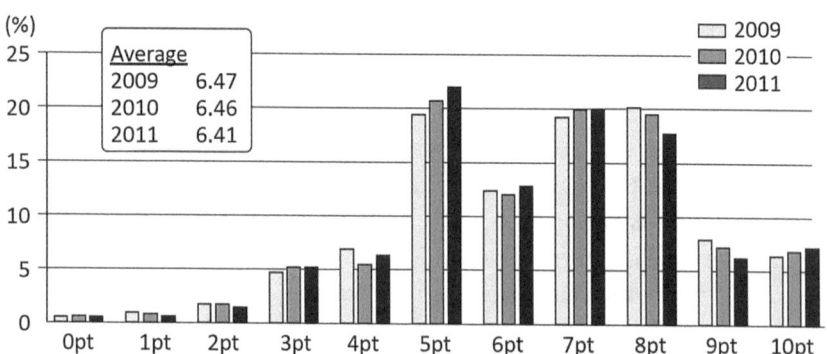

**Fig. 5.3**  Consistency of Life Satisfaction Scores in Japan over time (2009–2011)

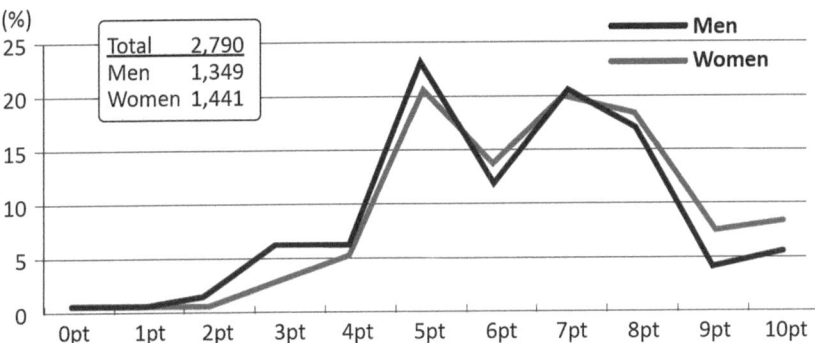

**Fig. 5.4** Consistency of Life Satisfaction Scores in Japan across sex/gender (2009–2011)

we see strong continuity. Although data is not yet available, we would expect to see the same continuity across the COVID pandemic as well. Figure 5.4 gives further indication of the cultural nature of responses: there is virtually no difference between the average male and female responses for these survey years.

We certainly recognize that there is much room for improvement in Japanese society. These include trying to achieve better work-life balance, inadequate time for family relationships and child care, and perhaps ways to raise hope about the future prospects of the society amidst Japan's prolonged recessionary economy. Recognizing differences should not become an excuse for non-improvement. However, as we have gestured to throughout, the models of happiness research proposed so far, and now institutionalized in global rankings, derive largely from the WEIRD North American context, and—taking one step further back—the European cultural, religious, and philosophical tradition from which it evolved. The questions included in these survey indicators/instruments already reflect the cultural values behind its creation and cannot be separated from them. Indicators, measurements, and rankings are all 'cultural products'.

In an earlier era defined by national borders and surveys (e.g., census surveys), we might assume 'fit', at least to some extent, between measurement and culture. Yet, the rise of *global* rankings that seek to encompass all cultures, force us to pause and re-assess. Indeed, the very assumption that a single global indicator might deliver a comparable view of happiness worldwide reflects a universalist assumption about happiness itself. Rankings require a single measurement criterion, but this in turn rest on a methodological assumption that myriad human communities worldwide are the same. Similarly, the inability to recognize larger cultural variations in happiness and well-being reveals a culturally determined predisposition to focus only on purportedly a-cultural individuals. Hypothetically, we might decide to measure happiness around the world by focusing on how much excitement people feel every day, to produce an index for the question "How much do you feel happy every day?". Yet, to do that would begin from the assumption that (1) excitement is an indicator of happiness, and (2) daily mood is a better indicator of happiness than a more long-term evaluation. Just to underscore this point, let us hypothesize an extreme example: What if Californians, where the average household has more than two cars and where public transportation is primitive, created an indicator that defined happiness as the number of cars a person owned, then measured it globally, and ranked countries accordingly. Would we accept this as a valid scale? The key point here is that cultural values and dispositions are inherent in all forms of measurement. By extension, these measurements function as culture products both reflecting the minds of those who make them, and influencing the minds of those who pay attention to them (policymakers, media, public, etc.).

What then are some of the assumptions inherent in existing indicators? One is, as touched on above, the notion of happiness as acquisition and attainment: 'getting' the life you want or a future 'ideal' in the Satisfaction with Life Scale (SWLS) and Cantril Ladder, respectively. A second assumption is the individual, an inward-reference point: 'I' get what I want, 'I' feel satisfied. Here the quality of relationships or surrounding environment that affords happiness is rendered invisible. Instead, focus is placed on the autonomy (or agency), inwardness, and motivation of the individual respondent. A third assumption, more prevalent among those who try to create hierarchical rankings, is that the higher the score, the more 'happy' a given group of people are. The Cantril Ladder was proposed by Hadley Cantril, a Princeton University psychologist, and despite working with the middle-class American population in the mid-1960s, his work claimed universalism with his title *The Pattern of Human Concerns* (1965). The Satisfaction with Life Scale (SWLS) was developed in the early 1980s by Ed Diener, a psychologist at the University of Illinois, whose subjects were 176 American undergraduates. Although derived from the North American cultural context, these WEIRD cultural products have gone global, carried forth by both the universal assumptions of their originators and the global shift to happiness.

Yet, Japan, and arguably most of East Asia, does not share these first assumptions. Instead of an internal, inwardly focused reference point, happiness in Japan is externally, outwardly focused. In line with deeply held ideas that the inner is indivisible from the outer environment (recall Nishida), judgments of happiness are gauged *in relation* to significant others and surrounding context. Second, in place of attainment, we find attunement: happiness signifies a resonance with the surrounding world; the degree to which one finds oneself in tune with the contextual milieu. Third, in a Yin-Yang conceptualization of happiness, the ideal range of happiness is the 'middle': a moderate level, with an awareness that extreme levels of happiness cannot continue indefinitely, and are likely to lead to extreme levels of unhappiness. In this mode, happiness is not an unalloyed good but—like all things—carries negative potential, for example, inviting jealousy or having adverse social consequences. Better to abide in moderation, in tune with surrounding others. Better not to get too carried away in the positive elements, blinding one to a latent negativity that will eventually reveal itself. Here, beyond mere cultural response bias, we can envisage how deeper cultural reasons play a role in Japan's 'low' marks on subjective well-being. Differences derive from deeper sources—such as self-construal and worldview—but are, when read through North American cultural products, understood as deficiencies.

In response to this situation, Uchida, alongside others in Japan, has developed an alternative measure: The Interdependent Happiness Scale (IHS). The IHS derives from assumptions closer to the relational, allo (other)-centric view of happiness dominant in Japan and much of East Asia. It measures individual perceptions of the interpersonally harmonized, quiescent, and ordinary nuances of the term. As shown in Fig. 5.5, sample items include 'I believe that I and those around me are happy', 'I

INSTRUCTION:
Q. Please indicate the degree to which the following statements accurately describe you using the scale from 1. Strongly disagree, 2. Somewhat disagree, 3. Neither agree nor disagree, 4. Somewhat agree, 5. Strongly agree. Please choose one option from below, and circle the number on the scale.

1. I believe that I and those around me are happy.
2. I feel that I am being positively evaluated by others around me.
3. I make significant others happy.
4. Although it is quite average, I live a stable life.
5. I do not have any major concerns or anxieties.
6. I can do what I want without causing problems for other people.
7. I believe that my life is just as happy as that of others around me.
8. I believe I have achieved the same standard of living as those around me.
9. I generally believe that things are going well for me in its own way as they are for others around me.

**Fig. 5.5** Interdependent Happiness Scale Items. (Adapted from Hitokoto & Uchida, 2015)

feel that I am being positively evaluated by others around me', and 'I can do what I want without causing problems for other people'.

Compared with SWLS, these items gesture toward and capture a different worldview and notion of happiness. Hitokoto and Uchida (2015) began developing the scale with Japanese college students, but then validated it—both among students and adults—in various different cultural contexts worldwide, including the USA, Germany, Korea, Thailand, Poland, and the Philippines (Hitokoto, 2014; see also Datu et al., 2016). This is a key point: it underscores that happiness, even in other cultural contexts, can contain the meaning of 'harmony with others'. In fact, as shown in Fig. 5.6, this scale tends to show less variation between diverse cultural groupings worldwide, as compared with those derived from life satisfaction.

That is, the gaps that emerge between Latin American countries, North America, and East Asia in the subjective surveys reviewed in Chap. 3 tend to be far less pronounced in this scale, strongly suggesting less cultural bias in the measurement instrument itself (recall the way the SPI, BLI,

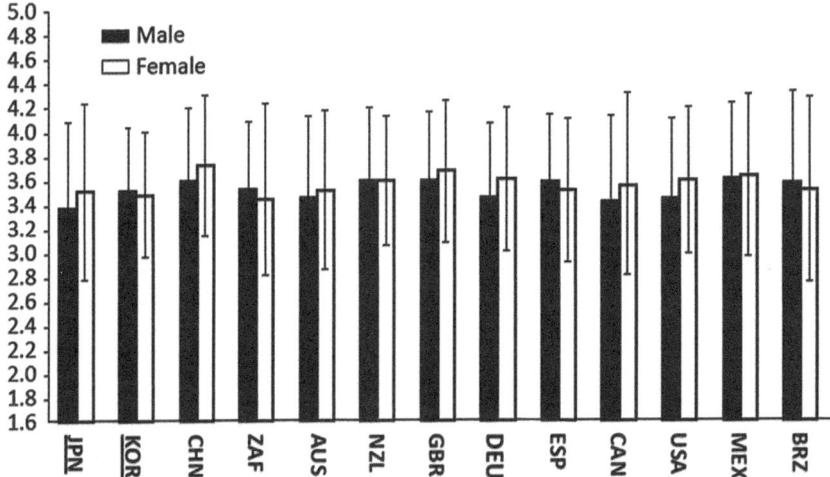

Fig. 5.6  Results of Interdependent Happiness Scale-based survey for select countries. (Derived from Koyasu (2012))

and WHR varied widely). We read this as evidence that an interdependent, attunement mode of happiness is *more widely understood worldwide*, as compared with an independent, attainment mode. Utilizing this alternative measure, we can confirm that happiness in Japan, Korea, and China is not in deficit compared with North America (USA, Canada) and Northern Europe (Switzerland). Different, not necessarily deficient.

Building on this work, Uchida has recently collaborated with leading researchers worldwide to produce the first Global Survey of Balance and Harmony (GSBH). The GSBH was incorporated into the 2022 World Happiness Report (WHR), demonstrating a successful case of the recognition of diverse modes of happiness. Introducing the GSBH, the authors of the WHR recognize a similar problématique to the one we have unfolded here:

Rather than only comparing cultures on concepts and metrics developed in Western contexts, there is increasing recognition of the importance of studying cultures through the prism of their own ideas and values, and of exploring cross-cultural differences in how people experience and interpret life. ... Arguably the most widely-studied cross-cultural dynamic is one that is germane to this chapter, namely the differences between Western and Eastern cultures. (WHR, 2022, p. 132)

The GSBH items included in the survey are shown in Fig. 5.7. Most of the items largely mirror the forms of happiness we have outlined thus far. That said, in the current volume we have not heavily discussed the notion of high-arousal versus low-arousal (excitement vs. calm) due to space. This is an important difference first recognized by Stanford psychologist Jeannie Tsai. Yet these items follow the contours of the discussion thus far: excitement signifies standing out, whereas calm represents attuning and fitting in.

The WHR 2022 chapter covering GSBH concludes by underscoring the universality of balance and harmony dimensions of happiness: "first, balance/harmony 'matter' to all people, including being experienced by, preferred by, and seemingly impactful for people, in a relatively universal way. Second, and relatedly, balance and feeling at peace with life could be considered central to well-being, on a par with other key variables linked to high life evaluations, such as income, absence of health problems, and having someone to count on in times of need" (p. 145).

- Balance: "In general, do you feel the various aspects of your life are in balance, or not?" [Response options: yes; no; don't know; refused to answer]

- Peace: "In genery do you feel at peace with your life, or not?" [Response options: yes; no; don't know; refused to answer]

- Calmness: "Did you experience the following feelings during a lot of the day yesterday?" [Followed by a series of feelings, including ...] "How about Calmness? [Response options: yes; no; don't know; refused to answer]

- Calmness preference: Would you rather live an exciting life or a calm life? [Response options: an exciting life; a calm life; both; neither: don't know; refused to answer]

- Self-other prioritisation: "Do you think people should focus more on taking care of themselves or on taking care of others?" [Response options: taking care of themselves; taking care of others: both; neither: don't know: refused to answer]

**Fig. 5.7**  GSBH survey items. (WHR, 2022)

To reiterate, instead of viewing happiness measurements and rankings as objective truth, it is better to see them as cultural products that can both distort and reveal, and inevitably play a role in reinforcing *or transforming* existing notions of happiness. These measurements can be distorting when the values underpinning the indicators are unrecognized, casting particular cultures (often those who created them) as leaders, and everyone else as deficient. Yet, the measurements, if done well, can also be revealing, bringing to light different dimensions of happiness and the human experience that are perhaps felt but not articulated in a given culture. We suggest that subjective well-being focused on independence and attainment tends, when exported globally, to distort happiness and constrain human experience, as it narrows the focus to 'internal attributes' alone. This sort of (mis) measure inevitably leads to further cultural products such as news stories, policies, and pedagogies that narrowly focus on individuals and subjective well-being, that is, news stories of Finland as happy followed by efforts to borrow their policies and pedagogies. Within the current context of the dominance of North American models, the diversification of models such as the GSBH serves to call attention to elements that are missing in the contemporary global discourse: balance, calmness, and allo-centric attention. In the same way, this recognition can lead, in turn, to alternative cultural products that place a greater emphasis on these elements. In terms of the earlier mutual constitution of cultures and selves conceptualization, these measures can help either reinforce or transform the selves that collectively constitute a society and culture. Transformation of the wider trend of global culture favoring WIERD approaches starts with a shift away from deficit to diversity.

# (Mis) Measurement, Institutions, and Practices: The Case of Education

In our modern societies, schooling has become the primary social institution charged with transmitting culture. Its explicit mission is to socialize and enculturate the next generation into the worldview of the previous one. Even in liberal systems, such as the United States, that claim to be educating youth to challenge the existing status quo by exercising *individual* agency, cultural continuity prevails. Moreover, it is to the education system that the adult world turns when it seeks to alter the status quo; when it attempts to reorient society toward new trajectories. As such, a focus on formal education (schooling) brings unmatched clarity to the themes at the heart of this chapter. It also adds a sense of urgency to the discussion: how we conceptualize and pursue student happiness today will, in large part, produce the world we live in tomorrow and the dominant form of happiness and well-being in the twenty-first century.

Within the wider global shift to happiness we have outlined, around 2015 the OECD's Education and Skills Directorate announced its departure from a narrow focus on academic skills toward happiness and well-being. Its overarching goal was announced as 'Well-Being 2030'. This partial pivot away from a narrow band of academic subjects—math, science, and reading—that purportedly indexed levels of 'human capital' in a given economy (country) was highly significant. The OECD was tracking the larger discoursal shift away from GDP, and built on the momentum of OECD's BLI launched several years earlier. Beginning in 2015, the PISA tests—the flagship OECD educational work, administered every three years across over 90 economies/countries worldwide—would add questions to its supplementary Student Questionnaire to gauge levels of student happiness. These results would be analyzed separately, and published under a stand-alone report entitled *Students' Well-Being* (OECD, 2017). In the PISA 2018 test, the OECD added further questions to understand the frequency of particular emotions in the lives of students. These included how often they felt 'happy', 'joyful', 'cheerful', as well as 'sad', 'miserable', and 'scared'. It also asked about students' 'meaning in life' by asking questions such as: 'My life has clear meaning and purpose' (Rappleye et al., 2023).

Figure 5.8 shows the results of the 2015 Student Well-Being Report (OECD, 2017, p. 71). The major East Asian countries are ranked at the very bottom: Japan, Korea, Taiwan, Macao, and Hong Kong, with China (four provinces) not far behind. At the top of the scale are countries of Latin America and the Caribbean: Dominican Republic, Mexico, Costa Rica, and Colombia. Clustered around the OECD average are many countries in Western Europe: France, Luxembourg, Germany, Spain, and

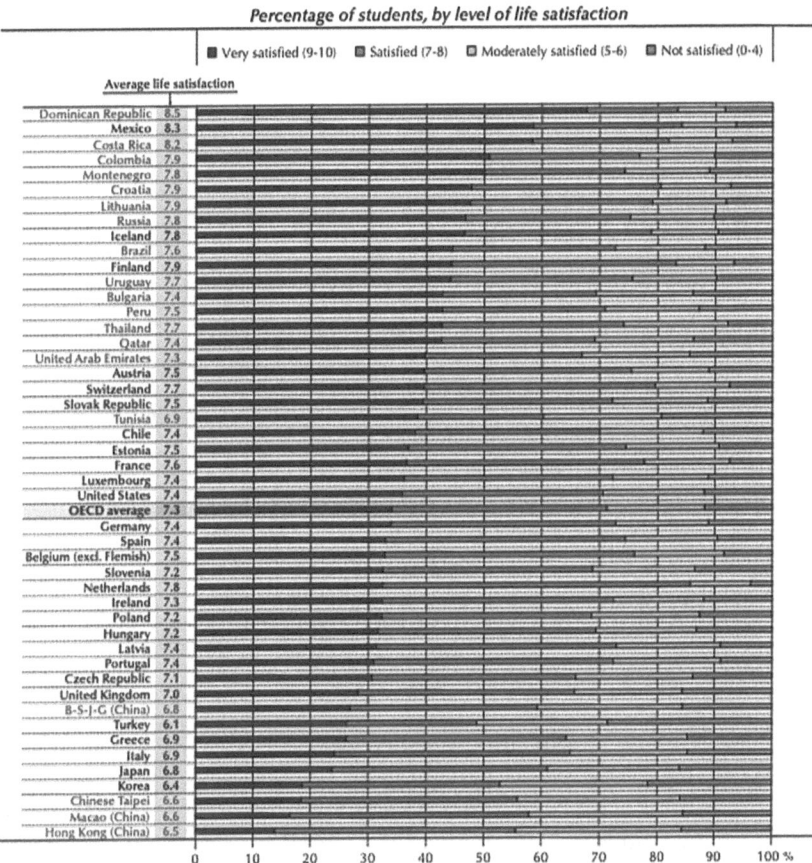

Fig. 5.8 OECD's survey of student well-being, results. (Adapted from OECD, 2017)

Belgium, with the United States also located here. Above average, but still below the leading Latin American countries are the countries of Finland, Russia, Lithuania, Iceland, and the Netherlands. Given the OECD's continued focus on student achievement, OECD analysts took the next step of correlating these well-being scores with student performance (i.e., PISA 2015 science scores), as shown in Fig. 5.9. Countries of northern Europe are in the 'High Satisfaction, High Performance' quadrant, and countries of East Asia in the 'Low Satisfaction, High Performance' domain. These results appear to confirm the WHR rankings. The results implicitly suggest that the education systems of Finland, Estonia,

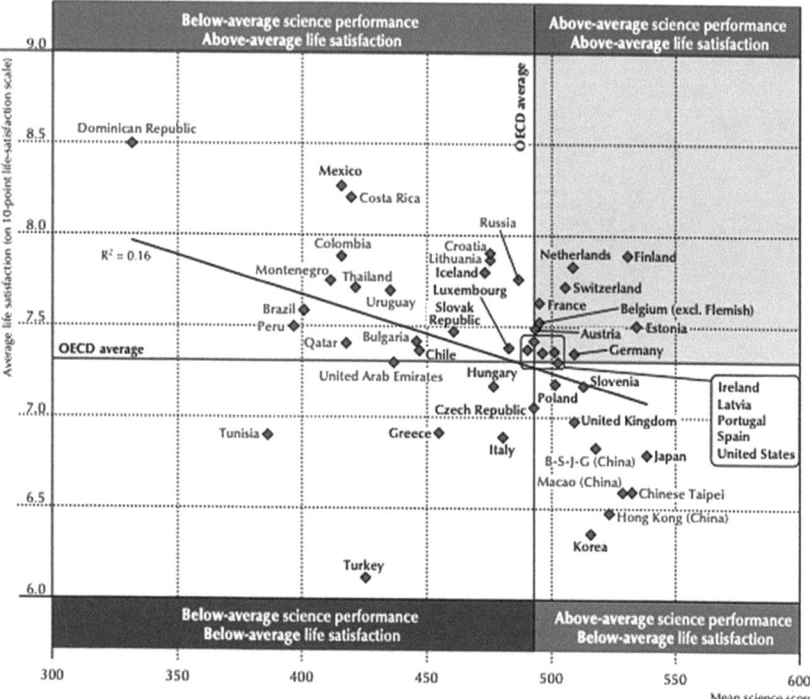

Fig. 5.9 OECD well-being scores and academic performance. (Adapted from OECD, 2017)

Switzerland, and the Netherlands have found a way to educate happy youth, whereas East Asian system were producing high-scoring, but unhappy students.

However, in light of previous chapters, how does the OECD's Education and Skills Directorate obtain its 'well-being' score in PISA 2015? The score comes *entirely* from student responses to the Cantril Ladder. In stark contrast to the OECD's larger BLI initiative that, as we saw in Chap. 2, included an expansive list of 11 categories and refused to rank, the OECD's Education analysts elected to equate student well-being *entirely* with subjective well-being (life satisfaction). This led, predictably, to the distortions we have pointed out earlier: East Asian countries score dismally. Perhaps the OECD education analysts were simply in search of a single, 'easy to use' measurement to compare countries globally, naively unaware of differences globally.

However, other evidence suggests that the OECD has been deeply and consistently committed to the independent mode of self. In launching the PISA initiative in the late 1990s, the OECD started formulating a set of Key Competencies, a normative framework to guide all future PISA studies and development of survey indicators. Here is a snapshot of those Competencies:

> Acting autonomously is particularly important in the modern world where each person's position is not as well-defined as was the case traditionally. Individuals need to create a personal identity in order to give their lives meaning....
>
> In general, autonomy requires an orientation towards the future and an awareness of one's environment, of social dynamics and of the roles one plays and wants to play. It assumes the possession of a sound self-concept and the ability to translate needs and wants into acts of will: decision, choice, and action. (Ibid.)

Beyond the valorization of 'autonomy' we see here, the same report then goes on to specifically admonish against an interdependent mode, as *individuals* 'need to develop independently an identity and to make choices rather than just follow the crowd' (OECD, 2005, p. 14). The report also explicitly mentions that individuals need to be 'optimistic' as they look to the future.

Despite the fact that Japan and Korea were full members of the OECD, no representatives from East Asia were invited to participate in the formulation of these Key Competencies or the 2015 PISA Well-Being Questionnaires. Instead, Western-trained cognitive psychologists—those working from a universalistic premise, and unaware of (or dismissive) of evidence developed by cultural psychologists—decided on the measures, that is, the exclusive use of the Cantril Ladder. This OECD's cultural product was subsequently adopted by leading global organizations like UNICEF. In 2020, UNICEF's *Worlds of Influence: Understanding What Shapes Child Well-Being in Rich Countries* report ranked Japan second to last in the category of 'Mental Well-Being', largely due to low subjective well-being.[1] These results led to a round of recrimination in Japanese media, and sparked debates within Japan's Central Council of Education around how to improve student happiness and well-being. We will come back to discuss this later development further below.

The OECD's assumption that schooling universally functions to create independent, autonomous, and acquisitive individuals is contradicted by decades of scholarship on East Asian education. Work on Japanese education alone is voluminous, and has consistently stressed the different aspects of the Japanese schooling system all converge on the notion of "close interpersonal relations, as the primary means for effective teaching and learning" (Shimahara & Sakai, 1995, p. 168). Captured in keywords such as *kizuna* (Shimahara & Sakai, 1995) and *minna* (Ueno et al., 2022) that are pervasive slogans across Japanese education, everything from official school goals and curricular guidelines to forms of pedagogy to micro-rituals in classrooms revolve around relations and attuning.

For example, goals for elementary school routinely involve statements such as 'get along well with others and help each other', while middle school textbooks utilize the metaphor of an orchestra to teach one's role in society: each instrument produces its own sound, but each finds its meaning in the larger ensemble. Pedagogical practices include, at the elementary level, small groups (*han*) as the primary unit and, at the high school level, whole-class teaching. Central to this teaching is a "pedagogy of feeling" (Hayashi et al., 2009), where affect is socialized in more allo-centric directions. Shared communal activities such as school cleaning and shared meals reinforce the idea that such modes are not limited to academic subjects, but an all-encompassing mode of living. The daily rituals of Morning Meeting (*asa no kai*) and Closing Meeting (*kaeri no kai*) found in homerooms even down to the preschool level, and collective bowing at the start

---

[1] We note that the UNICEF "Mental Well-Being" measure was a composite of the 2015 OECD PISA subjective well-being score and the youth suicide rate. Although the stereotypical image of Japan is that suicide rates are high, Rappleye and Komatsu (2020) have shown that they are around the average, less than the United States and far less than purportedly 'happy' places like Finland. That is, it was not the suicide rate, but instead the subjective well-being score that led to the low composite score.

of each lesson become the daily rituals that reconfirm the larger themes of relations and attunement. So central are relations to Japanese education, in fact, that some researchers point out that *kizuna* is "not a means to an end" but the end itself (see Shimahara & Sakai, 1995).

Contrary to the OECD's view, relations are not subsequently built between autonomous individuals, but are fundamentally constitutive of one's 'own' identity. 'Following the crowd' would be a gross misreading of what is taking place in these Japanese classrooms: the enculturation into a world of relations and attunement; socialization into an interdependent mode of self. And we insist this is not just something found in Japan: in Korea too the themes of interdependence and "affective relationality" are central (see Hyang, 2021).

We may briefly contrast this with the self that is 'schooled' in North America. So pervasive is the idea of individualism in the United States that it influences virtually all aspects of education:

- individualized instruction is the ideal, leading to smaller class sizes and ability grouping, as well as project-based learning driven by students' individual interests;
- a heavy emphasis on choice in classrooms and the wider curriculum (e.g., high school elective classes);
- a cadre of specialists to address the individual needs of students, including school counselors;
- Individual Education Plans (IEPs) for struggling students; Independent Study Contracts, often for talented students, "designed to respond to the pupil's unique educational needs, interests, aptitudes, and abilities" (California Department of Education, 2022);
- individualized, per-head funding schemes; an emphasis on self-direction, thinking for oneself;
- a system of college entrance applications that require a 'personal statement' where students lay out their unique path and attributes that make them worthy of admission.

And how about the affect modes enculturated in such schools? Growing up in California in the late 1980s and early 1990s, Rappleye attended elementary school at the zenith of the self-esteem movement. In 1986, a

California Congressman, concerned with rising crime and school under-achievement and building on educational scholarship valorizing the concept, launched the Task Force on Self-Esteem. The Task Force encouraged schools to provide opportunities for students to learn to evaluate themselves positively, focus more on accentuating the unique attributes of each individual student, and encouraging students to self-actualize. Indeed, the notion of self-esteem derives, in part, from Maslow's hierarchy of needs. But the California Self-Esteem movement suggested that, in changing the way students think about their situation, higher stages could be reached *sooner* with positive effects on behavior and spillover effects for society. Optimism was key. Students were all taught to be positive and focus on our individual 'potential'. The self-esteem movement led to a whole range of stand-alone programs that aimed to boost student self-esteem, self-expression, and self-worth, and led to a proliferation of other cultural products such as song lyrics, television shows, and, arguably, even social media paradigms such as California-based Facebook (Twenge & Campbell, 2009). Following movements like these, all across the United States, self-esteem is emphasized and this leads to children being taught from an early age to feel that they must have special, good, and unique traits (see Heine et al., 1999).

As discussed previously, North American research has identified the following as predictors of happiness: independence, control over one's surroundings, life goals, personal growth, and self-acceptance (Ryff & Keyes, 1995). It is largely this view that the OECD Education Directorate has taken up in its conceptualization of happiness, with the ideal being the autonomous individual, forging his/her own identity and path forward, and looking optimistically to the future. Using the Cantril Ladder, an instrument which derives from the same American milieu, the PISA empirical results seem to confirm the underlying conceptualization. Yet, is it really that East Asian students are unhappy? Or is it the case, perhaps, that the underlying conceptualization, measurement, and ranking work together to obfuscate alternative modes of happiness?

Unfortunately, we cannot get at this question directly, as the Interdependent Happiness Scale has yet to be added to global surveys on

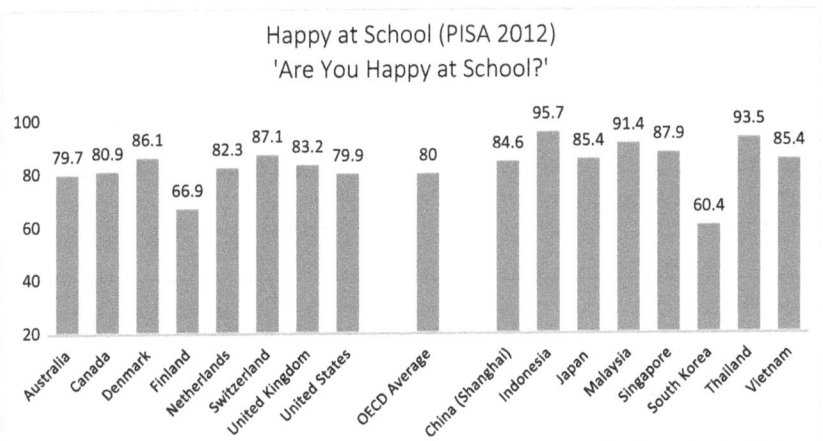

**Fig. 5.10** Happiness at school among select OECD countries (percentage of students who agree or strongly agree with the statement "I feel happy at school" in PISA 2012 (OECD, 2013, p. 32))

students. Yet, one way to get at the question is to look back to earlier OECD tests (PISA 2012) which asked a more open-ended question about happiness to students, simply: "Are you happy at school?" This question shifts the focus away from the individual, and more toward the place; away from individual mental states, and more toward a relational understanding of where the school sits emotionally in relation to other institutions in a given society. Figure 5.10 shows the results, focusing on East Asian countries, in comparison with Northern European countries that topped the PISA 2015 Well-Being ranking, and the North American countries that have formed our comparative reference point through this volume.

Here 85% of Japanese 15-year olds reported being happy at school. Other East Asian countries scored roughly the same, with the exception of Korea. Not only are these results above the overall OECD average, but these figures are roughly equal to the Northern European countries of Switzerland, Netherlands, Denmark, Sweden, and Belgium, with the exception of PISA-leaders Finland which scored below the OECD average. The United States scores below the OECD average, and well below the East Asian countries, including Japan and China. This calls into question the results of the individualized, self-esteem boosting approach found there. A range of recent research further corroborates that in Japanese schools, where relations and attunement are paramount, Japanese students, on the whole, feel less stress and pressure (Rappleye & Komatsu, 2018) and rates of bullying are lower, perhaps because relations are so heavily emphasized (Rappleye & Komatsu, 2020), as compared with American students. One study from 2017 that looked at student's subjective feelings in greater detail found that East Asian high school students, on the whole, have far more positive feelings about school than their US counterparts, as shown in Fig. 5.11 (see Komatsu & Rappleye, 2020).

Our point here is that these 'alternative' measurements and indicators provide much evidence to support an argument of a different form of happiness at work in East Asian schools. It is not, as the OECD's rankings suggest, that East Asia is in deficit, but likely that its interdependent

| Question | Japan | US. | China | Korea |
|---|---|---|---|---|
| **Goal** | | | | |
| One of your goals is to get into a well-known university (%) | 49.9 | 82.5 | 81.4 | 68.0 |
| One of your goals is to be of a high social standing (%) | 48.9 | 71.4 | 75.9 | 63.0 |
| Study-related anxiety (%) | | | | |
| I want to be absent from school (%) | 32.6 | 48.2 | 39.3 | 33.9 |
| I want to quit school (%) | 8.0 | 35.1 | 17.4 | 21.9 |
| I want to skip class (%) | 27.8 | 35.3 | 18.3 | 28.7 |
| I cannot sleep (%) | 17.3 | 48.6 | 36.3 | 11.8 |
| I feel depressed (%) | 46.8 | 32.4 | 71.7 | 39.2 |
| I feel irritated (%) | 49.9 | 66.2 | 68.6 | 43.4 |
| I feel inferior (%) | 37.2 | 24.3 | 32.9 | 39.4 |
| I want to destroy things (%) | 9.0 | 17.3 | 19.8 | 14.2 |
| I want to blame someone or scream (%) | 5.2 | 19.7 | 26.5 | 6.5 |
| I despair in my life (%) | 12.3 | 13.6 | 15.3 | 22.3 |
| Others (sorts of anxiety) (%) | 3.4 | 10.3 | 3.1 | 3.4 |
| I have never felt or experienced any of the above (%) | 14.6 | 11.1 | 7.7 | 12.8 |

[a]The survey method is described in NIYE (2017) and Rappleye and Komatsu (2018).

**Fig. 5.11** Results of comparative survey of 10–12 graders for four countries. (NIYE, 2017)

modes of happiness are not being captured within the dominant Western conceptualization and measures. East Asian students are being taught the interdependent mode daily, and they are learning their lessons well. Nonetheless, they are cast as deficient globally by leading organizations such as OECD and UNICEF.

It is worth noting that Japan's Central Council for Education (JCCE), the highest policy advisory organ in the country, has recently focused heavily on student happiness and well-being. Yet, it is now well aware of the mismatch between the Western metrics and culturally dominant forms of happiness existing in Japan. Uchida was elected as a member of the Central Council in 2020 and was requested to help in the formulation of new, more appropriate indicators. Together, the authors—Uchida and Rappleye—have given numerous talks to the Council, and leading policymakers in the Ministry of Education as they seek to develop a culturally appropriate response to the wider global shift. Indeed, several of the ideas and examples found in the current volume were first shared in the wider policy discussions within Japan (see Uchida & Rappleye, 2022).

# (Mis) Measurement, Institutions, and Policies: Regional Happiness, Organizational Culture, and Social Capital

Earlier we shared Dewey's observation that the Protestant legacy led to 'individual capitalism', a position congruent with the sociological work of Max Weber. Today, most economists continue to promote these individualism-centered capitalist economic models, largely unaware of the underlying cultural assumptions. Take, for example, the idea of Human Capital, the theory that the knowledge and skills of individuals aggregate to drive organizational success and national economic competitiveness. Recently, the World Bank launched the Human Capital Project, formulating a Human Capital Index that seeks to calculate the potential economic productivity of individuals born in a given country over the span of their working life. The Bank subsequently called attention to a 'Global Learning

Crisis', wherein education is portrayed as failing to generate the requisite human capital necessary for the benefit of a given society (World Bank, 2018). The assumption is, again, that individuals acquire knowledge and skills and, in the aggregate, these improve the lives of a given country. This basic framework is, in turn, resonant with a socio-political model that emphasizes individual rights, freedoms, and choices, and seeks to keep institutional constraints loose. This model gives rise to the free market, a socio-economic model in which each individual pursues his/her own interests, organizations compete for talent, and national competitiveness is paramount. All of this is pursued under the belief that the 'invisible hand' of the free market will bring the greatest happiness to the greatest number. At one point in the 1980s, the belief in this model was so strong that the UK's Prime Minister Margaret Thatcher claimed there is "no such thing as society", only individuals.

However, even a cursory review of the past few decades of free market globalization reveals the limits of such thinking. A company competing with rivals attempts to reduce the cost of its product, and thus makes the decision to shift manufacturing overseas where labor costs are cheaper. The result is that cities/communities rapidly lose employment when a factory moves. A network of secondary industry and services is lost alongside it. The working-age population leaves the local area to look for work and, as a result, the local area experiences a decline in income. This has been a common and devastating drama playing out across Japan, and increasingly across Korea and Taiwan. In low-income developing countries such as, say, Nepal and the Philippines, workers do not simply migrate to major cities, but instead leave their country altogether to work overseas in places like the Middle East. These workers send remittances back home, and these are recorded as economic gains for the country as a whole. Yet, the impact on local collectives—families, communities, and workplaces—is stark. Or take deregulation. A policy strategy promising greater efficiency in market allocation of human capital, deregulation has led to an increase in the use of temporary staff and short-term workers rather than regular employment. As a result, workers have less money in their wallets, and less rootedness and connection to a given organization and community.

American sociologist Robert Putnam (1995) famously put forth the notion of Social Capital, shifting the focus away from human capital. Defined as "the features of social organization such as networks, norms, and social trust that facilitate coordination and cooperation for mutual benefit" (p. 35), it called attention to the ways *relations*—coordination, trust, and cooperation—lead to benefits. For Putnam, the focus was on how the rapid decline in America's social capital would endanger American democratic institutions. His work now appears prescient. But in the context of our volume we may extend this idea, by combining it with the interdependent mode we have been describing, in order to rethink what we should invest in now within the shift away from the twentieth-century GDP=Happiness equation.

For most economists, shifting to think about happiness and well-being, the nation or national society remains the focal point. Meanwhile, most psychologists tend to focus on the individual. Yet, when we examine our everyday lives we interact, identify, and feel allegiance to smaller units, such as family, school, workplace, community, or region. It is these meso-level units that we tend to feel and recognize the interdependent modes we have been describing here. If one feels happy skipping off to work one morning, but arrives to find oneself surrounded by unhappy co-workers, happiness and well-being are hard to maintain. In fact, organizations and companies in Japan have long been aware of this, and tend to engage in a large number of 'collective activities'—practices that may seem odd or intrusive from a Western corporate perspective. In Japanese companies, people work together in units called 'islands' or 'lines'. It is common to see office layouts where the desks face each other and the direct supervisor can look over their subordinates. They often wear similar uniforms. Events such as drinking parties, company trips, athletic meets, and morning meetings (sometimes involving shared cleaning and calisthenics), which were customary in virtually all Japanese companies in the past, are also collaborative. Although the 'thickness' of these intra-organizational relations vary according to type of industry, job level within the company, and even where the company is located (Tokyo tends to adopt more modern, Western work styles), it is widely recognized that building such 'connections' within an organization plays a role in reducing loneliness and improving employees' sense of well-being.

Sharing of organizational culture (i.e., explicit organizational philosophy) and collaboration between the company and the community are also thought to have a positive effect on the mental well-being of employees. Based on this, it is important that social scientists and psychologists alike begin to pay greater attention to the 'middle' (meso) level of organizational culture. That is, an interdependent mode of happiness necessarily focuses on collectives over individuals, putting the policy emphasis on investing in the social capital of the meso-units.

More concretely, in contemporary Japan, rural communities are acutely aware of and concerned about the problems generated by a rapidly declining population (aging, combined with outward migration of youth to cities in search of work). In many areas there is a sense of resignation, a feeling that nothing can be done. Although the older people wish to preserve the villages, their children have already left to the cities. In such cases, simply measuring levels of happiness is insufficient. Creating new measures, specific to these regional and local communities is important. Capturing the diversity of regional, community, and organizational units within measures becomes necessary. How do residents of a given location feel about immigration from other areas? What unique traditions or natural sights bring a sense of happiness to local residents? Such conversations at the meso-level are important for creating the conditions for interdependence to flourish. Based on these discussions, policies can be created that maintain connection and share values within those regions, communities, and organizations (such as strengthening and mobilization around key cultural festivals, in the case of Japan). Creating such policy discussions and guidelines helps engage people in re-evaluating the social, natural, and cultural environment in which they live and understanding how these link to collective happiness and well-being.

Given that the current volume is primarily focused on the global discussion and most readers will be less familiar with the specificities of localities in Japan, we have chosen to devote less space to the discussion of the meso- and local aspects. Yet, in other work, we have engaged in substantial research and developed models looking at this dimension (e.g., Uchida & Takemura, 2012). As shown in Fig. 5.12, we envisage the investment in meso- and regional social capital as catalytic for a virtuous cycle, leading to higher levels of interdependent happiness alongside the

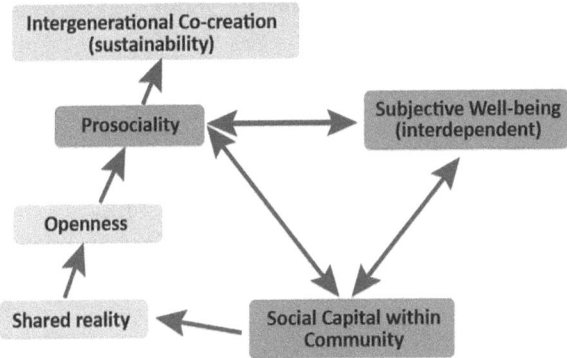

**Fig. 5.12** Social capital and sustainability: an interdependent approach

continuation of traditions and creative re-articulation of those interdependent ways of living. Amidst the push to think about happiness globally, it is important for countries such as Japan, Korea, and Taiwan, which have all experienced decades of emphasis on 'individual capitalism' and GDP=Happiness, to remember that interdependent happiness is fostered more within the intimate meso-spaces between nation and individual.

# References

Cabinet Office, Commission on Measuring Well-Being. (2011). Measuring National Well-Being—Proposed Well-being Indicators. Available in English Online at: https://www5.cao.go.jp/keizai2/koufukudo/pdf/koufukudosian_english.pdf

California Department of Education. (2022). Independent Study. California Department of Education. https://www.cde.ca.gov/sp/eo/is/

Datu, J., Valdez, J., & King, R. (2016). The Successful Life of Gritty Students: Grit Leads to Optimal Educational and Well-Being Outcomes in a Collectivist Context. In R. King & A. Bernardo (Eds.), *The Psychology of Asian Learners* (pp. 503–516). Springer.

Dewey, J. (1930, February 19). Individualism: Old and New. New Republic. https://vdoc.pub/documents/individualism-old-and-new-1mss4u3fpqbo

Hayashi, A., Karasawa, M., & Tobin, J. (2009). The Japanese Preschool's Pedagogy of Feeling: Cultural Strategies for Supporting Young Children's Emotional Development. *Ethos, 37*(1), 32–49.

Heine, S., Lehman, D., Markus, H., & Kitayama, S. (1999). Is There a Universal Need for Positive Self Regard? *Psychological Review, 104*(6), 766–794.

Hitokoto, H. (2014). *Interdependent Happiness: Cultural Happiness under the East Asian Cultural Mandate.* JICA Research Institute.

Hitokoto, H., & Uchida, Y. (2015). Interdependent Happiness: Theoretical Importance and Measurement Validity. *Journal of Happiness Studies, 16*, 211–239.

Hyang, J. (2021). Individualization, Peer Learning and the Cultural Model of Sociality in South Korean Education: The Case of an Educational Metaphor. *Ethos, 49*(19), 51–73.

Kitayama, S., Park, H., Sevincer, T., Karasawa, M., & Uskul, A. K. (2009). A cultural task analysis of implicit independence: Comparing North America, Western Europe, and East Asia. *Journal of Personality and Social Psychology, 97*, 236–255.

Komatsu, H., & Rappleye, J. (2020). Reimagining Modern Education: Contributions from Modern Japanese Philosophy and Practice? *ECNU Review of Education, 3*(1), 20–45.

Koyasu, et al. (2012). Kofukukan no kokusaihikaku kennkyuu—13 ka koku no deta [International Comparisons of Happiness—Data from 13 Countries]. *shinrigaku hyoron* (Japanese Psychological Review), *55*, 70–89.

Markus, H. R., & Kitayama, S. (2010). Cultures and Selves: A Cycle of Mutual Constitution. *Perspectives on Psychological Science, 5*, 420–430.

Markus, H., Uchida, Y., Omoregie, H., Townsend, S., & Kitayama, S. (2006). Going for the Gold: Models of Agency in Japanese and American Contexts. *Psychological Science, 17*(2), 103–112.

Mauss, M. (1898 [2000]). *The Gift: The Form and Reason for Exchange in Archaic Societies.* W.W. Norton.

Morling, B., & Lamoreaux, M. (2008). Measuring Culture Outside the Head: A Meta-analysis of Individualism—Collectivism in Cultural Products. *Personality and Social Psychology Review, 12*(3), 199–221.

National Institute for Youth Education. (2017). Koukouseino benkyoto seikat-sunikansuru ishikichosa houkokusho: Nihon, beikoku, chugoku, kankokuno hikaku [A Report for Learning Practice, Belief, and Attitudes of Upper Secondary School Students: Comparison among Japan, the United States, China, and Korea] [in Japanese]. National Institute for Youth Education. http://www.niye.go.jp/kenkyu_houkoku/contents/detail/i/114/

Nishida, K. (1932). Jitsuzon no konteitoshite jinkaku gainen [The Concept of Personality as the Foundation of Reality (Lecture Delivered at the Shinano Education Center from 3–5 September)]. In N. Kitaro (2020). Collection of Lectures by Nishida Kitaro. Iwanami Press (pp. 19–32).

OECD. 2005. *The Definition and Selection of Key Competencies: Executive Summary*. OECD. Retrieved June 19, 2018, from https://www.oecd.org/pisa/35070367.pdf

OECD. (2013). *PISA 2012 Results: Ready to Learn: Students' Engagement, Drive and Self-Beliefs*. 3 vols. OECD Publishing.

OECD. (2017). *PISA 2015 Results (Volume III): Students' Well-Being*. OECD Publishing.

Putnam, R. (1995). Bowling Alone: America's Declining Social Capital. *Journal of Democracy, 6*(1), 65–78.

Rappleye, J., & Komatsu, H. (2018). Stereotypes as Anglo-American Exam Ritual? Comparisons of Students' Exam Anxiety in East Asia, America, Australia, and the United Kingdom. *Oxford Review of Education, 44*(6), 730–754.

Rappleye, J., & Komatsu, H. (2020). Is Bullying and Suicide a Problem for East Asia's Schools? *Evidence from TIMSS and PISA. Discourse, 41*(2), 310–331.

Rappleye, J., Komatsu, H., Uchida, Y., Tsai, J., & Markus, H. (2023). The OECD's 'Well-being 2030' Agenda: How PISA 2018's Affective Turn Gets Lost in Translation. Comparative Education (in press).

Ryff, C. D., & Keyes, C. L. M. (1995). The Structure of Psychological Well-Being Revisited. *Journal of Personality and Social Psychology, 69*(4), 719–727.

Shimahara, N., & Sakai, A. (1995). *Learning to Teach in Two Cultures: Japan and the United States*. Garland.

Shweder, R. (1991). *Thinking through Cultures. Expeditions in Cultural Psychology*. Harvard University Press.

Twenge, J. M., & Campbell, W. K. (2009). *The narcissism epidemic: Living in the age of entitlement*. Simon and Schuster.

Uchida, Y., & Rappleye, J. (2022, July). Kyouiku seisaku no well-being [*Well-Being in Education Policy*]. Presentation to Japan's Central Council for Education, Tokyo, Japan (Delivered Online). Presentation Slides Here. https://www.mext.go.jp/kaigisiryo/content/000177757.pdf

Uchida, Y., & Takemura, K. (2012). No wo tsunagu shigoto: fukyuushidouiin to comunitei he no shakaishin rigakuteki apuroa-chi [*Working on social ties: Social psychological approach toward extension officers in agricultural communities*]. Soshinsha.

Ueno, M., Fujii, K., & Kashigawi, Y. (2022). Philosophy of Minna and Moral Education: Manabi That Encompasses Everyone. *Education Philosophy and Theory (Online First)*. https://doi.org/10.1080/00131857.2022.2094243

World Bank. (2018). *LEARNING to Realize Education's Promise (World Development Report)*. The World Bank.

World Happiness Report. (2022). *The World Happiness Report*. UNSDNS. https://worldhappiness.report/archive/

# 6

# Interdependence: Alternative for the Twenty-First Century?

The last two chapters have elaborated the interdependent approach, first conceptualizing it, then relating examples of its cultural manifestations. We now turn to examine the potential significance of the interdependent mode for contemporary problems we—*collectively*—face at the global level. Instead of viewing the interdependent mode as a mere empirical descriptor of happiness and well-being across East Asia, in this chapter we gesture to its potential import globally. The crux of our argument is that WEIRD globalization has placed a heavy burden on contemporary youth, encouraging forms of subjectivity, development, and well-being that are difficult, if not impossible, to sustain in the contemporary economic and environmental climate. In the search for alternatives that can respond to these challenges—most of all, the sustainability imperative—we present emerging evidence that underscores the potential of the interdependent approach, not just for East Asia but globally. In this way, we advance the discussion from 'alternatives to us' to 'alternatives for us' (Geertz, 1973), resisting a relativist argument in favor of a pragmatic search for new solutions to shared problems. We also address temporal change in this chapter, resisting an a-historical, culturally essentialist reading of interdependence. Cultural change is constantly unfolding, and our role—at least as we see it—is to attempt to shape that change in a pragmatically useful direction in the face of an uncertain twenty-first century.

© The Author(s) 2024
Y. Uchida, J. Rappleye, *An Interdependent Approach to Happiness and Well-Being*,
https://doi.org/10.1007/978-3-031-26260-9_6

# New Models for the Twenty-First Century? Globalization, Sustainability, and the Independent Mode

> Now more than ever, the need for a different development approach is highlighted in ecological, social, and economic crises: ecosystem degradation, potentially catastrophic climate change, excessive consumption of the affluent and extreme poverty on the other end, and growing inequalities both between and within nations. Underlying all these crises is the lack of a holistic view that would focus on causes instead of symptoms, and *the inadequacy of the architecture of global governance to address these problems.* ... To properly assess well-being outcomes, a more integrated measurement system that balances the ecological, social and economic and cultural dimensions of development is needed. (New Development Paradigm Initiative, 2014, p. V11, italics added)

As reviewed in previous chapters, the seismic discoursal and policy shift toward happiness and well-being unfolding over the past decade has largely been driven by a loss of faith in the GDP=Happiness equation of the twentieth century. The 2011 UN resolution, *Happiness: Towards a Holistic Approach to Development*, explicitly raised the call for a new paradigm. In its wake followed work such as the New Development Paradigm Initiative, cited above. In 2015, the optimism of progress found in the Millennium Development Goals (MDGs) officially gave way to the far more sober Sustainable Development Goals (SDGs). Thus, within the space of a single decade, economic, social, and intellectual paradigms that seemed so certain in the twentieth century were buckling under growing evidence that contemporary models were no longer sustainable.

Fascinatingly, however, deep reflection on the 'contents' of this new model/paradigm did not accompany this shift. Instead of a sole focus on economic growth, happiness and well-being would now take center stage. Yes. But what forms of well-being? What modes of happiness? As reviewed in the opening chapter, many of the new global indices of happiness, including the World Happiness Report (WHR) and the 2015 PISA Student Well-Being studies aimed to address the SDGs, simply *assumed* the universality of the independent approach dominant in the Protestant

West, and began measuring all the world by this standard. Was there empirical evidence to support the idea that this familiar independent mode would lead to a 'different development approach'? Or was it an old wine, new bottle scenario? Were these measures created unreflectively, and without any sense of alternatives? Can we expect that 'cultural products' emerging from the same cultures from which the previous unsustainable paradigm emerged will affect a new model? One key issue is, as pointed out in the quote above, the "inadequacy of the architecture of global governance" to recognize, let alone incorporate, alternatives. This lack of alternatives comes back around, in the model of the culture cycle discussed in the last chapter, to reinforce, on a worldwide scale, those non-sustainable ways of (well-)being.

Forming the larger backdrop here is the past few decades of globalization. One version of the globalization story focuses on the economy, production, and technology: out-sourcing, off-shore manufacturing, and communications advances have led to the global integration of markets, and upon this economic base we find increasing cultural and social globalization unfolding as well. This materialist version of globalization issues from a similar perspective as twentieth-century GDP-ism. Another version of the story is that contemporary globalization represents not simply the triumph of liberal market systems, but a much more expansive set of psycho-cultural pressures on non-Western countries; that is, the spatialization of Western modernity and post-modernity. In many countries, including Japan and much of East Asia, these cultural aspects of globalization are highlighted, with globalization frequently carrying the less felicitous sense of unwelcome 'Westernization' or 'Americanization'. Cultural products like global happiness rankings and, say, educational 'best practices' promoted by the OECD, UNESCO, and UNICEF are not inherently aimed at furthering the market economy, but do advance this psycho-cultural dimension.

A core element of this "psycho-cultural globalization" (Jung & Ahn, 2021) is the spread of Western-style individualism. According to one study that analyzed the cultural transition in Japan and the USA on the axis of individualism-collectivism between 1950 and 2008, a period of rapid economic development in both countries, there has been a common increase in individualism in both contexts, as measured by a decrease in the number of household members, an increase in nuclear families or unmarried persons living in urban areas, and an increase in rates of divorce (Hamamura, 2012). Here we see that economic changes underpin, in some ways, social and thus cultural changes. We do not intend to deny this fact. Yet, the idea that similar economic changes produce or require similar psycho-cultural change is too simplistic. In the 1960s, Geert Hofstede attempted to quantify countries along an Individualism-Collectivism continuum, an important precursor to much of the cultural psychology work of today. One would expect, following a materialist assumption, that economic change leads directly to cultural transformation, that East Asian countries would—following their explosive economic expansion in the second half of the twentieth century—be much more individualistic today. However, Minkov et al. (2017) have recently reexamined the Hofstede values with data collected in 2014–2016, and confirms that East Asian countries are still not highly individualistic.

Let us again turn to the Japanese case to understand these dynamics better. Undoubtedly, in Japan a market-based economy has put pressure on a culturally embedded interdependent mode of self-construal and well-being. As discussed in the previous chapter, Japanese corporations had traditionally placed a heavy emphasis on interdependence, and this was institutionally manifest in a range of shared activities and lifetime employment schemes ('the salary man' image that is so well-known abroad). Yet, the rise of contract (non-salaried) positions, labor fluidity, and corporate restructuring have forced Japanese workers to act more like independent individuals: working to ensure their own security and future prospects, prioritizing an individual career path over the needs of the corporation, and so on. Unfortunately, the decline in organizations predicated on an interdependent mode has led many Japanese to understand individualism, or individualistic ways of working, as a sort of denial of relationships. This has led to 'isolationism' as the means of achieving individualism. Many readers will have heard of the problem of *hikikomori* (literally: those who withdraw), wherein working-age Japanese adults refuse to enter society. Nakatani (2008) shows how the spread of neoliberal values and institutions has led to the loss of security and the foundation of trust between people in Japanese society, producing these sorts of problems.

Thus, while one can marshal some evidence in support of a narrative that globalization has erased alternatives, we see a more complex picture. The actual impacts of globalization on changing Japanese values are somewhat superficial. The underlying approach of collective orientation and interdependent modes remain strong. It is precisely for that reason that we see conflicts and challenges arising. For some, the interdependent mode appears to be under threat. Many Japanese are no longer allowed or encouraged to seek a sort of quiescent happiness shared with others. Even if they would rather pursue cooperation or attunements, they are being encouraged and institutionally incentivized to seek competition and self-assertion. As Japanese companies become more performance-based and competitive, it is difficult for individuals to completely distance themselves from the competitive reality, even if they are not interested in competition for advancement. Thus, we see that in contemporary Japan, there are two ways of being, an individual independence mode that gets progressively stronger at the discoursal level, and a relational interdependent mode that remains strong in spheres less touched by globalization.

Unfortunately, there are many instances where frameworks for understanding these differences excessively emphasize the opposition between the two. Take, for example, the Commission for the Design of 21st Century Japan (1999). It was written in the late 1990s by an influential group of Japanese political leaders, just as the impacts of globalization were beginning to be strongly felt:

> Unfortunately, Japanese society still frowns on displays of individual excellence. This is closely bound up with an ingrained egalitarianism bordering on the absolute. … The tendency of the Japanese to regard the harmony of their immediate surroundings as paramount has had the merit of creating a nation with minimal disparities in wealth and a high degree of safety relative to other developed countries. But instead of letting individuals give full rein to their abilities and creativity, these social settings have turned into shackles. (p. 8)

This Report goes on to suggest that a lack of 'robust individuality' has become the prime impediment to Japan's economic resurgence globally, continuing:

In the 21st Century ... Japanese will be required to assert themselves as individuals and to possess a robust individuality. The kind of individual called for at this time is, above all, one who acts freely and with self-responsibility, self-reliantly supporting himself. This 'strong yet flexible individual' takes risks self-responsibly and tackles the challenge of achieving personal goals with a pioneering spirit. (p. 8)

As we see here, under the continued influence of GDP-ism and buoyed by global (read: Western) discourses, there is a tendency to wholly dismiss an interdependent model, despite its recognized benefits (e.g., less inequality, higher degree of safety), and instead place sole faith in 'robust individuality'. The latter quote here is strikingly similar to the OECD's Key Competencies reviewed in the last chapter. At the same time, the contrast with the New Development Paradigm is stark.

In contrast to the simplicity of these sorts of policy discourses, we believe that fostering happiness and well-being in contemporary Japan begins with recognizing the interdependent mode, and then searching for ways to support a version of interdependence capable of weathering the tide of contemporary globalization. Our own view of the contemporary Japanese psycho-cultural situation is a two-story house, as depicted in Fig. 6.1. The ground floor is an interdependent mode, and the second floor is an independent mode. If the interdependent mode is the foundation, rooted in religio-philosophical narratives spanning thousands of years, the second floor has only just been added to the Japanese psyche, beginning with Westernization during the Meiji Restoration (1868) and accelerating greatly in the past few decades of neo-liberal globalization that emphasizes "individual freedom", and manifest in the cultural forms of neo-liberal policymaking and economic reforms. We note that a second floor can be easily expanded, rearranged, and redecorated. But the first floor, on the other hand, is an indispensable part of the building's entire structure. Compared to lavish discoursal decorations over the past few decades taking place on the second floor, the first floor may appear non-descript and plain. Yet, there is no way to do anything on the second floor in Japan without having first entered the interdependent first floor.

The second floor houses the independence pursued by the discourses of 'free competition' or 'global values', and is where the furniture of

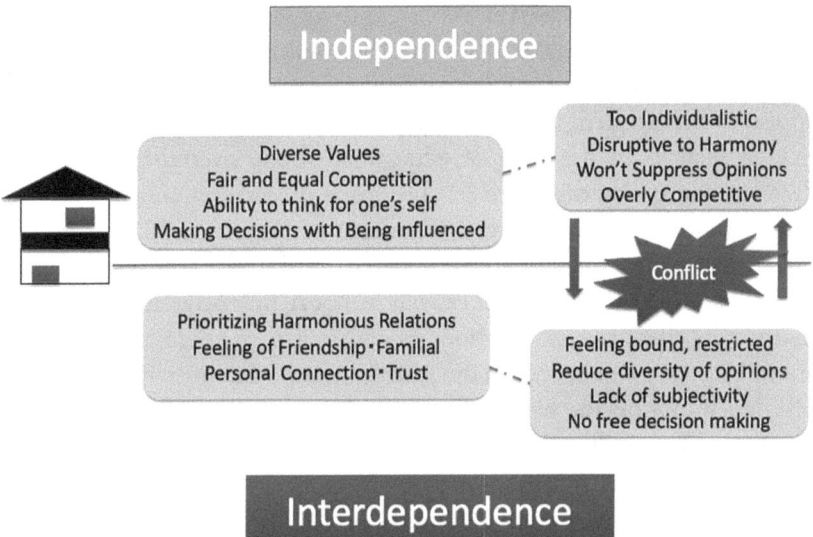

**Fig. 6.1**  'Two Story' conceptualization of modern subjectivity, case of Japan

'individual freedom', 'individual rights', and 'uniqueness' sit. Yet no matter how well elaborated, independence in Japan is an 'afterthought' on the second floor, not the foundation. Both individual freedom and the desire for uniqueness—core concepts in neo-liberal economics and education for self-esteem alike—are difficult to achieve without a deep understanding of the atomized, independent mode of self that predominates in the West. However, the acquisition of an independent 'cognitive frame' has only been accumulated, particularly in the Protestant United States, through a long history of religious beliefs and Western philosophical thought. When Japanese try to adopt only a part of this way of being 'individual', they do so within a context where the culture, secular philosophies of modernity (e.g., Nishida philosophy) and social institutions (e.g., education system) do not support this discourse. As such, distortion and conflict inevitably occur.

In fact, in the most globalized spheres of Japanese society, such as the business world (think: Toyota) and policymaking emphasizing economic growth (think: Ministry of Finance), the first floor of internal harmony has been roundly dismissed as a conservative and troublesome old 'habit'. It is viewed as something 'backwards'. These sentiments come out strongly in the quote from the *Commission* above. But the same sentiments also appear in the frequent dismissal of Japanese philosophical ideas and theories, as well as in the field of education where, over the past several decades, the policy discourse has been one of derision and the need to create 'global human resources' (Komatsu & Rappleye, 2021). What is lost in this rather unfortunate and largely uninformed political discourse is the way that interdependence may support independence. Although many view these two orientations as contradictory, there are arguments about how they can be made compatible, hybridized, or switched 'on and off' depending on the needs of a given situation (see Kasulis, 1998). To over simplify, a weakening sense of trust on the first floor creates structural problems for independence on the second floor. In one empirical study, we conducted on local communities in Japan, we found that towns with high levels of trust within the community were more 'open' in their attitude toward accepting newcomers and change (Uchida et al., in prep). At first glance, we might expect that trust within a town would lead to exclusivity, as if 'only people from this town are trustworthy'. However, we found that when there is greater trust with others, communities are able to work together to deal with challenges. This includes welcoming newcomers who bring new ideas. Also, where relationships of trust predicated on interdependence exist, people are able to evaluate each other fairly without fear of misunderstanding. In other words, the cooperative nature of the first floor could be compatible with the independence of the second floor if it is used as a system for building and maintaining mutual trust, rather than a conservative and hierarchical one.

Our rather extended discussion of Japan is important for three reasons. First, recognizing change underscores that we do not see cultural patterns as essential and unchanging. These cultural patterns are constantly evolving, which is the very reason we need to think carefully about which patterns lead to which futures, and decide upon which patterns we seek to support. Patterns of culture are held in place, often only delicately, by the sorts of discussions and institutions we create and engage with. Second, a view of Japan's changes under globalization brings into focus, we believe, a dynamic unfolding across most East Asian societies. Given that globalization in the independent mode is now common to all these countries, but these same countries simultaneously lack the deep Western religio-philosophical roots of individualism, what is happening in Japan may be a guide to thinking about what is happening across East Asia, and perhaps elsewhere in the world. Third, this discussion of Japan brings into focus why interdependent modes of happiness and the operationalization of those modes in forms such as, say, the interdependent Happiness Scale are crucial. As countries, like Japan, come to be increasingly impacted by global discourses, it is essential that those discourses are diversified enough to encompass and support modes of (well-)being found there. If global change is not inclusive, then it is merely hegemonic. Kitayama and Markus (2000) wrote more than a decade ago that "often as innocuous and well-intended as they are, various attempts to apply theories of happiness that are implicitly grounded in Western ideas of progress, liberalism, egalitarianism, and freedom to other cultural contexts may not reveal but distort lived experience of the people in those cultures". We wholeheartedly agree, adding that when such theories become the basis for policymaking, this goes beyond mere distortion: it accelerates the elimination of alternatives. These other ways of (well-)being may well offer—as we shall see below—more effective solutions to twenty-first-century challenges.

## The Transition Generation: Youth and Educating the Future of Well-Being

Before moving to that larger discussion, it is worth spotlighting the youth and the cultural arena of education. These have been a consistent theme for us throughout this volume, as the youth and their forms of education reveal our collective future, already unfolding. The breakdown of twentieth-century models is most acutely felt among the youth, many of whom hold a vision not simply of breakdown but of bankruptcy. Numerous polls show a growing cynicism with twentieth-century policies, as the gap between the political rhetoric of optimism clashes with the pessimistic socio-economic realities that youth find themselves are faced with. Let us again focus on Japanese youth and their education, then extend that discussion globally.

Are Japanese young people happy or not? The data on subjective happiness is, in fact, unclear: some results show that levels of happiness are generally declining among the youth, while others show that it is increasing. In the next section we will look closer at the latter data, as it is somewhat counter-intuitive but also highly instructive. What is clear, however, is that the objective economic and social conditions that Japanese youth find themselves in have not changed in a positive direction. The rate of full-time employment is decreasing, the number of part-time employees is rising, and young people's anxiety about the future has grown considerably, as manifest in, say, a consistently declining rate of marriage. Following the acceleration of globalization in the late 1990s, discourse and policies aimed at promoting competition (competition in a market-based economy) have become stronger in Japan. The country—once among the most powerful economies in the world—has been exposed to severe price competition by rising 'rivals' across East Asia. As a result, costs associated with human resources have been greatly reduced, working styles diversified, and, inevitably, inequalities have risen (Yamada, 2009). Exacerbating all of this, Japan has experienced a prolonged economic slump after the collapse of the so-called Bubble Economy in the early 1990s. Within Japan, the past few decades are often referred to as the Lost Decade(s). Originally it referred to one decade, now it is going on three.

Faced with this, Japanese companies, which have an enduring seniority-based system, responded to these challenges in ways that maintained a relative continuity in conditions for mid-career workers (middle management) and above, but shifted dramatically the patterns of hiring and employment among young people: unstable contract posts, reduced benefits, and simply a refusal to hire youth (Genda, 2001; Toivonen et al., 2011). In 2010, while Japan's overall unemployment rate for all ages was 5%, the unemployment rate for those aged 25–34 was 6.2%, and the number of part-time workers exceeded 30% of the total (Ministry of Health, Labor and Welfare, 2013). In addition, the job offer rate for college graduates (as of December of the year before graduation) has been consistently falling since 1997. The rate did bottom out in March 2011

(68.8%) and has been rising again. But few remain optimistic about the future. Moreover, the rate of young people leaving the workforce within three years is also extremely high, suggesting that the quality of the jobs on offer entail poor conditions or are underpaid: about one-third of all new graduates quit their jobs at an early stage. The lucky ones moved to better jobs. The unlucky ones ended up unemployed or simply in 'withdrawal'.

Under such changing conditions, shifts in the view of happiness among Japanese youth are inevitably undergoing change. Compared to their parents or grandparents generation (anyone over 50 years old), who grew up at a time of strong economic growth, the goal of 'work hard and get rich' is no longer a strong goal among Japanese youth. In fact, compared to young people around the world, Japan ranks the lowest in terms of the desire to "earn more money than my parents" (Zielenziger, 2007). This data is corroborated by a 2010 survey conducted by the Japan Youth Research Institute comparing the views and attitude of high school students in Japan, South Korea, China, and the United States. In response to the question, "Do you want to be a great person?", the percentage of students who answered "strongly agree" or "agree" was lowest among Japanese youth: 86% for China, 72% for South Korea, and 66% for the USA, but just 43% for Japan. In the previous chapter, we saw that only 49% of Japanese youth had the goal of achieving a high social standing, the lowest among these countries. We read these results as suggesting that, although Japanese young people are in a difficult situation economically, they are more inclined to maintain the status quo and have close relationships with their surroundings than to 'chase dreams and aspire to greatness' as the postwar Japanese generation did. Instead of focusing on a material-rich future, one that looks increasingly unlikely, they instead seem to focus now on 'present happiness'. That is, Japanese youth look to be increasingly detaching happiness from the GDPism of the past, allowing happiness to increase amidst difficult socio-economic conditions (see Komatsu, Rappleye, & Uchida, 2022a).

Evidence corroborates this reading. Figure 6.2 shows data from the Public Opinion Poll on Citizens' Life conducted by the Japanese government annually since 1948, focusing on the last two decades (1999–2009). Here we see that the youth are actually reporting higher levels of

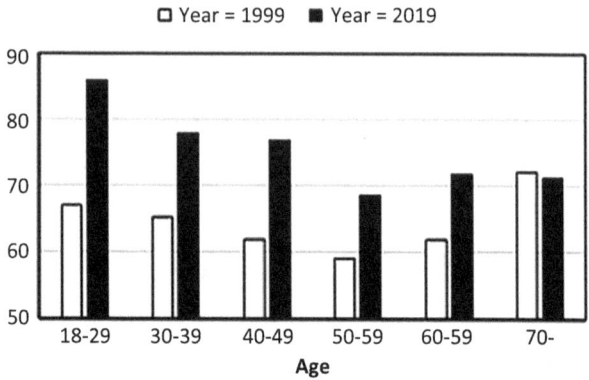

**Fig. 6.2**   Happiness among different ages: change across two decades in Japan

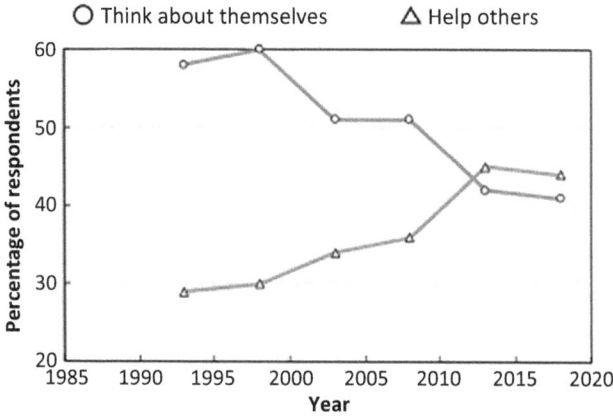

**Fig. 6.3**   Changing beliefs about self and others over the last three decades in Japan

happiness than older generations. There seems to be, among Japanese youth, a growing sense of 'kinship' and their preference for a slow life, a vision of young people not ambitious but who find happiness in taking care of their surroundings (Furuichi, 2011). As shown in Fig. 6.3, another survey shows that the percentage of respondents (all ages) who believe most people primarily think about themselves has declined, being overtaken by the percentage who believe that most people prefer to help others. This data suggests a movement away from wealth=happiness equation, momentum toward a new definition of happiness that focuses on relations, in particular 'safe' and stable interpersonal relationships. That is, there is a return to harmony in the present context, rather than a strengthening focus on independent and individual acquisition in the future.

We must, of course, be cautious in reading the data this way, as we may sound like apologists for an increasingly unfair socio-economic system. Indeed, we have already pointed out the negative dimensions of this transition, centered on those who cannot cope and simply withdraw (*hikikomori*) or refuse to participate in education and employment (NEET). In terms of *hikikomori*, we read this behavior as the plight of youth who have fallen out of the relational, interdependent arena(s) that define Japanese society. In 2010, it was estimated that there were 70,000 people who had exhibited withdrawal from society by the age of 39, and a 2009 survey suggested another 60,000 'middle-aged withdrawn people' at age 40 or older. Unfortunately, once these *hikikomori* have 'fallen out' it is very difficult for them to return. We have conducted further research into the reasons youth are prone to these options, as seen in the development of the "NEET and Hikikomori Risk Scale" (Uchida & Norasakkunkit, 2015). Utilizing it, we found three types of orientations to this sort of withdrawal: (1) an attitude that necessarily rejects working in a hierarchical society with traditional Japanese norms, (2) a lack of confidence in one's communication and social skills, and (3) lack of clarity or uncertainty around the 'future goals' one wishes to pursue. Those at particular risk of becoming *hikikomori* are those who deviate from or reject the wider Japanese-style interdependence orientation, have a lower sense of well-being, and have fewer close relationships in their communities. It is frequently pointed out that Japanese youth seem to have declining motivation to try new things, and when asked what motivates them, they often reply that they are motivated only by what they like to do and/or are good at, but not by anything else. This lack of motivation connects, it seems, to the lack of clarity or uncertainty around future goals as well.

While not naïve to the negative dimensions of this transition, it is important to recognize the emergence of new forms of happiness among youth, as they may be pointing us to a different world, one more apropos for the twenty-first century. The danger is that policymaking, led by an older generation still entrenched in the twentieth-century views, reinforces the very outlooks that contribute to youth unhappiness and turn a blind eye to these alternative approaches.

As an illustrative example, let us briefly examine the recent movement to improve "self-esteem" among Japanese students. In the mid-2000s, amidst the policy discourses of individualism reviewed above and neo-liberal globalization celebrating the individual, several surveys emerged showing low levels of self-esteem (*jiko kotei kan*) among Japanese students. Pushing to one side the issues of response bias and cultural differences we have focused on throughout this volume, many prefectural Boards of Education (local education authorities) across Japan responded to these surveys by actively promoting pedagogies aimed at raising 'self-esteem'. For example, one prefecture in the middle of Japan distributed materials to all its teachers, explaining Maslow's Hierarchy (self-esteem purportedly linking to Maslow's Level 4) and promoting lessons that "encouraged students to recognize their individualism". This included the ability to 'make individual decisions' and 'set one's own goals'. Kyoto Prefecture, where we live, distributed guidelines to all local elementary schools encouraging lessons that asked students to "imagine what person they want to become in the future" (*naritai jibun*), and asking teachers to change the way they speak to students. Teachers were encouraged to put emphasis on positive words, individualization, and individual strengths, as opposed to pointing out negative aspects, whole-group discussion, and flagging individual weaknesses. At the same time, a range of popular books aimed at teachers emerged, sharing ideas about lessons that could improve self-esteem. One popular volume entitled, 'The Shower of Individual Praise', encouraged teachers to spend time in each lesson conveying the strengths of each student, so that each individual student emerged from the shower of praise "shining" (Kikuike, 2015).

While these new pedagogies are well-intentioned, the problems here are numerous. First, as we have seen, the notion of individuals with high levels of self-esteem is a particular cultural arrangement of North America. At root, it derives from an independent form of self-construal and happiness at odds with the Japanese context. Among policymakers, at least some, there is a continued misreading of difference as deficit. Second, based on this misreading, policies and practices are introduced that, while seemingly alleviating the deficit, actually accelerate the move toward independent modes. Making 'individual decisions' and imagining who one will be in the future sounds very much like the OECD's Key Competencies and the Cantril Ladder. Nonetheless, lessons that end with a 'Shower of Individual Praise' sit within an educational system that is deeply committed to fostering an interdependent mode. Japanese students then face mixed messages: become a 'strong individual' but learn attunement; think first of oneself (ego-centric) but stay committed to responsiveness to the other (allo-centric) forms. In a more critical appraisal, one could argue that the traditional forms of meaning and value—an interdependence mode—are being actively discouraged (e.g., not praising the overall efforts of the class).

In the surveys cited above, we have seen how there has been a shift in emphasis, away from pursuing one's own goals and toward 'helping others'. We have seen how Japanese youth appear to be increasingly satisfied with a life less focused on the pursuit of wealth and social standing. Our own empirical studies, conducted among college students and adults in Japan and the United States, found that in Japan, people who emphasized individualistic tendencies tended to be less happy (Ogihara & Uchida, 2014). The effect was mediated by the number of close friends: the stronger the individual achievement orientation, the harder it is for people to connect with others, and this may be a factor in lowering happiness. This same effect was not found in the United States. Here we see that the policymaking-turned-practice discourse is not only blind to these differences, but encourages attitudes that may actually lead to greater unhappiness. Japanese companies and schools have, under psychosocial globalization, shifted toward independent achievement orientations, but these systems are not undergirded by the personal values and perspectives that govern these systems in the North American cultural context. In this sense, the independent mode may be more difficult for Japanese, and other East Asians, to adopt, and subsequently lead to a range of unintended negative effects. We wonder aloud: instead of encouraging a transition to independent modes of happiness, shouldn't the emphasis be on repairing the interdependent mode threatened by changes in the wider socio-economic structures under neo-liberal globalization?

Here the larger picture comes into view. Despite a different dominant pattern of happiness at play in East Asia, global comparisons conducted within the narrow range of Western theories of happiness and well-being suggest East Asia to be in deficit. This deficit view leads to the introduction of new cultural practices that promise improvement, but (1) are at odds with the socio-economic context, and (2) block from view alternative forms of happiness. This much has been established already. But the key point is that in uncritically accepting the 'global' diagnosis, the next move becomes uncritically adopting practices from the more 'advanced' countries. Scant attention is paid to alternative practices, precisely because they are not well represented in the existing cultural products of global rankings, academic theory, or pedagogical practice. Yet, purportedly 'advanced' practices were developed at a very different time (i.e., periods of high economic growth), and in very different contexts (i.e., Protestant cultural sphere). Moreover, even if these practices 'improve' subjective happiness and well-being to some extent, youth with higher levels of self-esteem and independent self-construal are unlikely to be able to pursue the "person they want to become in the future" (*naritai jibun*), given declining resources and the breakdown of twentieth-century models. What is the way out of this?

Without being naïve to the negative dimensions, one way is to follow the youth themselves: reconnecting with an alternative sense of happiness and well-being, finding practices that support those modes, and thus shifting to a focus on the 'place' of happiness rather than individual disposition.

# Sustainability, Disaster, and Collective Action: Interdependence as Alternative Approach?

Among the challenges that humans collectively face, perhaps none is more pressing than the sustainability imperative. The *New Development Paradigm* (2014) cited at the outset draws attention to "potentially catastrophic climate change", a call taken up globally via the somber Sustainable Development Goals (SDGs). The youth, those destined to face the consequences, were—at least until the COVID-19 pandemic—protesting in the streets worldwide in the *School Strikes for Climate* movement, demanding immediate action on the climate crisis. Recall the Japanese government's 2011 conceptualization of well-being led toward 'sustainability' (Fig. 3.8). Even the OECD, an organization dedicated to furthering capitalist modes of economic growth following the Second World War, recently argued in the *OECD Environmental Outlook to 2050: The Consequences of Inaction* (2018) that environmental sustainability must become the foremost policy priority: "Humanity has witnessed unprecedented growth and prosperity in the past decades. ... This growth, however, has been accompanied by environmental pollution and natural resource depletion. The current growth model ... could ultimately undermine human development."

However, it is precisely 'inaction' that has defined the debate so far. Despite decades of scientific evidence, global agreements, and economic incentives such as carbon trading schemes, humans have collectively been unable to change course. Building on work done with our close colleague Hikaru Komatsu, we thus suggest cultural change as an alternative approach to addressing the crisis. The crux of our argument is that the North American independence model seems to fit an expansionary, growth period without resource constraints, whereas an interdependent model may be a better fit with a degrowth society, one defined by severe resource constraints. That is, when resources are expanding, maximizing individual gains may lead to growth and efficiency. However, in a degrowth scenario, where the amount of new resources that can be acquired is limited, rather than seeking to maximize individual happiness, sustainability must be redefined as collectively shared and/or in pursuit of 'moderate' levels of happiness.

Some recent work in the emerging field of environmental psychology has already prepared the way for this cultural approach. Arnocky et al. (2007) focus on attitudes toward the environment, reporting that individuals with an independent self tend to show only ego-centric concern (i.e., concern about environmental degradation because of the negative impact it will have on oneself) instead of eco-centric concern (i.e., concern about environmental degradation because humans are a part of nature). Other studies find that individuals in the independent modes are less effective in controlling their desires for the sake of social and ecological improvement (Martinsson et al., 2012; Chuang et al., 2016). One consequence of this is that those with independent self-construal tend to engage in pro-environmental actions with less frequency, for example, sorting garbage and driving less (Chuang et al., 2016; Davis & Stroink, 2016).

Our own recent work has attempted to go beyond merely differences in attitudes and pro-environmental behaviors, to instead examine actual impacts on carbon dioxide emissions and resource depletion (Komatsu et al., 2019, 2020, 2021; Komatsu, Rappleye, & Silova, 2022b). Utilizing the measure of ecological footprint, we found that countries where the dominant form of self is independent tend to have a higher ecological footprint, as shown in Fig. 6.4 (Komatsu et al., 2019).

In related work, and here connecting to the theme of education, we have also conducted studies finding that among high-income countries where independent self-construal dominates, forms of pedagogy such as student-centered learning also tend to dominate (Komatsu et al., 2021). Yet, it is these independent-heavy societies that are the least sustainable,

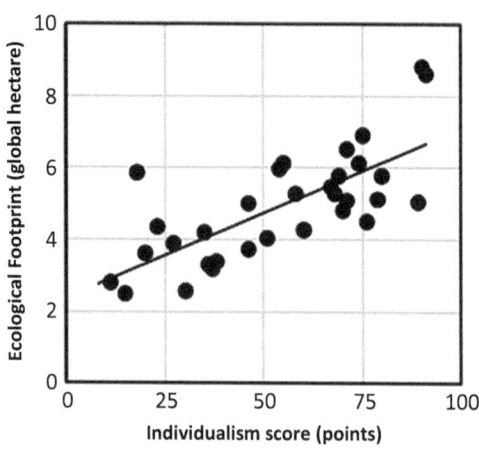

**Fig. 6.4** Relationship between individualism scores and ecological footprint of consumption. (Adapted from Komatsu, Rappleye, & Silova, 2022b)

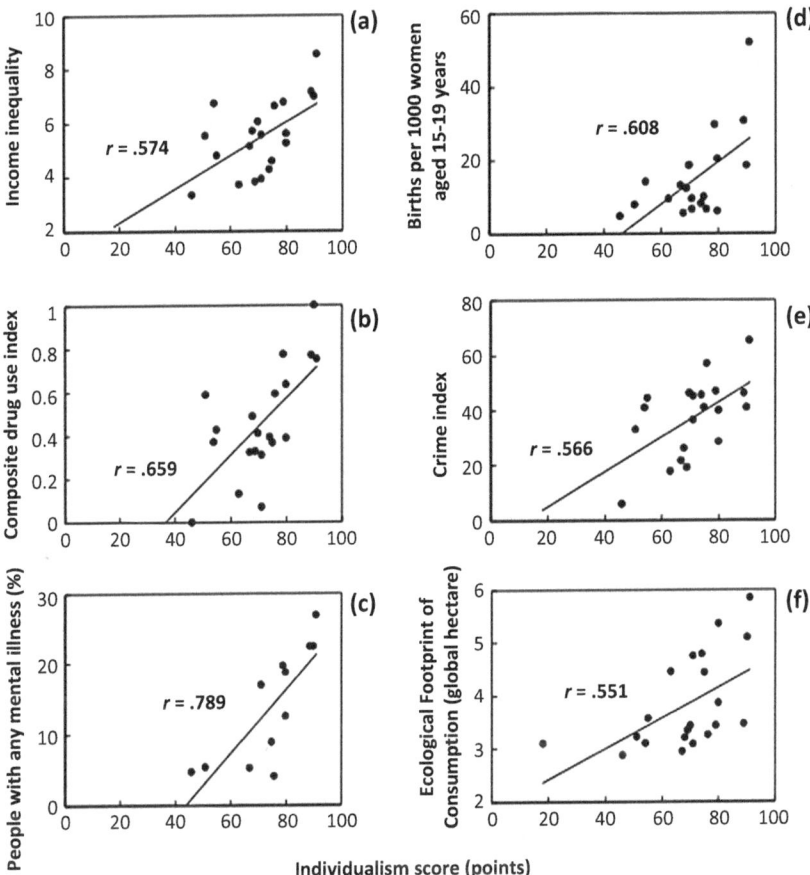

**Fig. 6.5** Relationship between independence (individualism) and various sustainability indices. (Adapted from Komatsu et al., 2021)

both environmental and across a range of social indicators, including income inequality, drug use, mental well-being, crime/safety, and stable families, as shown in Figs. 6.5.

While the results for social (un)sustainability are interesting enough, we wish to maintain a focus on environmental sustainability. This same study found that countries with high scores for independent self-construal (individualism), including the United States, UK, and Australia, tend to favor liberal market economies, and yet these economies show higher levels of ecological footprint of consumption. Japan—which, alongside Korea, has the lowest levels of individualism among high-income OECD countries—has a lower ecological footprint. Interestingly, Japan has seen substantial reductions in consumption-based environmental indicators over the past decade (15% in per capita $CO_2$ emissions, 18% in per capita ecological footprint; 27% for material footprint) despite the lack of strong government control (Komatsu, Rappleye, & Uchida, 2022a). This is the same period in which, as we reviewed previously, Japanese youth were transitioning to new forms of happiness, one more defined by interdependence and moderation.

At both a macro-global level and country-level comparison then, there is emerging evidence to support the notion that self-construal affects environmental attitudes and impact. Coupled with the failure of other approaches to environmental sustainability, we wonder aloud whether it is time to open the climate discussion to culture. More specifically, is it time to raise awareness around the interdependent mode within these discussions?

Models centered on the individual pursuit of happiness tightly couple with capitalist models: vigorous competition purportedly leads to higher achievement among individuals, in turn generating benefits for society as a whole. Whether or not this is actually the case (recall the Easterlin Paradox), the precondition for this model is that competition among individuals does not bring about a state of co-poverty due to damage to the macro environment. We find that the dominant models of self, happiness, and economy arose in a period of European history marked by expansion and under a cultural assumption of infinite resources (Rappleye & Komatsu, 2020). Although we are now aware that resources are finite, we continue to assume models of self, happiness, and economy developed at that earlier time. Even the major United Nations agencies leading the SDGs, including UNESCO whose mandate is education and culture, have largely failed to recognize the dimension of culture, let alone different modes of self-construal (see Komatsu et al., 2020).

As an alternative, the proliferation of 'cultural products' that promoted an interdependent outlook would contribute to changes in self-construal, and—over time—likely contribute to sustainability. Cultural change is not as elusive or impossible as we might imagine. Studies in cultural psychology have highlighted the notion of 'priming': simply reading a story with an interdependent theme or even circling interdependent pronounces (e.g., we) instead of independent pronouns (I) in a word search task increased interdependent self-construal (Brewer & Gardner, 1996; Gardner et al., 1999; Trafimow et al., 1991; see also Nisbett, 2003). 'Cultural products' are, in a sense, part of the priming landscape that impacts self-construal. Yet, what we see globally, despite all the rhetoric of sustainability and the purported shift away from GDPism, there continues to be a proliferation of cultural products that reinforce the independence model of the expansionary period, for example, all the happiness rankings we listed in the opening chapter and the pedagogical models scattered throughout. Can we really expect sustainable change to arise when the underlying cultural priming and cultural products remain unchanged?

## But Isn't It Already Too Late…?

More pessimistic readers will no doubt argue that the odds are now overwhelming that human society will fail to achieve environmental sustainability. It is therefore either idealistic or irrelevant to continue discussing ways to 'achieve' sustainability. We actually agree. Current countermeasures to environmental problems, including climate change, are not commensurate with the rate and magnitude of the changes predicted. Even if all the countries implement the policies promised in the Paris Agreement framework, global $CO_2$ emissions are projected to *increase* not decrease (Nieto et al., 2018; United Nations, 2021). Today, it may no longer be useful to talk about sustainability but instead *survivability*. Indeed, Kyoto University, where both of us have worked for a decade or more, recently established the School of Human Survivability, arguably the first in the world. It disposes with the pretense that sustainability is still a viable option. Yet even if we openly acknowledge that the current trajectory is unsustainable and pivot to prepare for the disasters to come, we still insist that the interdependent mode is crucial. Societies dominated by independent modes are expected to have far greater difficulty in adapting to the inevitable consequences of the looming crisis. This point has been confirmed by research in quite disparate fields, including social and environmental psychology, disaster science, and adaptation science.

For example, several studies in social and environmental psychology found that an independent self was less effective in controlling one's own desire for the sake of interdependent-collective social benefit (Seeley & Gardner, 2003; Chuang et al., 2016). Another study in environmental psychology (Arnocky et al., 2007) reported that an independent self cooperated less effectively with others than interdependent selves under hypothetical conditions of resource constraints. In the emerging fields of disaster science and adaptation science, recent work has found that societies with weak social cohesion and great inequality tended to be slow in recovering from catastrophic conditions induced by disasters (Dynes, 2006). Furthermore, in communities with weak social cohesion a higher percentage of people suffer from mental distress such as post-traumatic stress disorder after disasters. In one in-depth study, comparing the aftermath of Japan's Fukushima nuclear (2011) disaster with the Hurricane Katrina (2005), it was found that resilience was far higher in Japan, given far higher rates of social capital in Japan. We tend to view, in line with the argument in the last chapter, the relationship between interdependent modes and social capital as inseparable, and mutually reinforcing.

Indeed, the Fukushima disaster of March 2011 in which a tsunami and earthquake led to nuclear meltdown revealed many things about Japan. It helps us hypothetically envisage the future. In Japan, where major earthquakes and typhoons are frequent, there is an acute sense of how disaster can have an immeasurably large impact on the human mind and shape culture, thus making this a particularly robust field of research there. In terms of Fukushima, one survey conducted in the Tohoku region before and after the Great East Japan Earthquake in January 2011 and February 2012 found, somewhat unsurprisingly, that subjective well-being after the disaster was lower than before the disaster, and that this tendency was particularly strong in the major disaster-affected prefectures (Iwate, Miyagi, and Fukushima) (Horige, 2013). Another survey conducted in June 2011, three months after the earthquake, showed that the tendency of post-traumatic stress was significantly higher in the disaster-affected areas than in the unaffected areas, indicating the need for medical support and clinical counseling (Kyutoku et al., 2012;

Kotozaki & Kawashima, 2012). It is perhaps obvious that natural disasters lower happiness and well-being, but even man-made disasters have negative impacts on happiness, even among those not significantly impacted by actual events. For example, the September 2001 terrorist attacks in the United States had a significant impact on people outside of the USA such as in the UK, where it was reported that feelings of happiness declined after the attacks (Metcalfe et al., 2011). In addition, according to a monthly survey of adults across the USA about Hurricane Katrina from August to October 2005, negative emotions "felt during the week" increased in early September, the month when Katrina's damage became most apparent (Kimball et al., 2006). It seems obvious to us that the climate change will lead to more frequent disasters, and these will become an increasing drag on subjective happiness in coming decades, even for those of us not directly affected. Some work is already pointing to 'climate depression' as a new affliction, one particularly strong among the youth (Kalmus, 2021).

In Japan, one of the most interesting findings to emerge from the field of disaster research is the way that such events tend to strengthen the interdependent mode. Following the Great Kobe Earthquake (1995), one study sought to understand the psychological changes among university students by conducting a survey four to seven years after the disaster (Nishimoto & Inoue, 2004). It found that when faced with the threat of nature, there was an increase in the importance of connections with others and the appreciation of family and friends. Similar results were confirmed after the Fukushima disaster. For example, researchers at Keio University found that altruism increases in the aftermath of a disaster, as manifest in donations and increasing number of people involved in activities aimed at helping others (Ishino et al., 2012). This finding resonates with data from another public opinion poll: the percentage of people who believed that most people think about themselves reduced greatly after the Fukushima disaster, whereas the percentage of people who believed that most people want to help others increased (Komatsu, Rappleye, & Silova, 2022b). The latest round of the same survey conducted in 2018 also confirmed a comparably high percentage of people who believe that most people want to help others.

Notably, a first survey for the Cabinet Office's Happiness Index we featured in Chapter Three took place in December 2010, just three months before the disaster struck. Obviously there was no way to know that the Fukushima disaster would unfold a few months later. It was a large-scale happiness survey of 20,000 young people in their 20s and 30s. As a panel survey, it was designed from the outset to have the same person answer twice. The second survey was scheduled to be conducted at the end of March 2011. But since the second survey came just after the earthquake, the decision was made not to include residents of the six prefectures most affected by the disaster. New participants were recruited to replace those who had been left out. Analyzing this data in light of the Fukushima disaster, Uchida et al. (Uchida et al., 2014) found that: (1) As Kimball et al. (2006) pointed out, people felt depressed after the earthquake, so temporary positive emotions decreased and negative emotions increased; (2) at the same time, the experience of the earthquake changed people's sense of values, and they began to reevaluate their environment and the existence of others, which they had taken for granted; and as a result, (3) their criteria for judging happiness changed, and happiness tended to increase.

The second (later) survey included a question about change, added in light of the disaster: "Has your way of thinking about life and happiness changed?" In response to this question, a total of 58% of the respondents answered "greatly changed" or "somewhat changed", indicating that more than half of the respondents had experienced some form of change in their outlook on life and values. Those in this group also responded affirmatively to the direction of such change: "emphasis on connection", "emphasis on individual effort", and "feeling of emptiness". Overall, the change in emphasis on connection was the highest. These results show that more than half of the young people in their 20s and 30s experienced some kind of change in their outlook on life and values after Fukushima, even though they did not live in the immediate disaster zone. In the wake of the disaster, the Japanese word 'kizuna'—meaning 'fundamental connection'—became more prominent.

If what happened in Japan is any indication, a collective future marked by frequent climate-related disasters is likely to encourage a deep rethinking of what happiness and well-being mean. This rethinking will, of course, occur with greatest intensity among those directly affected by such disasters, as material acquisition and individual gain proves to be impossible. But more generally, we expect to see shifts among those not directly or intensely affected, as well. We find it hard to imagine how future disasters can be linked to arguments in favor of greater independence. Instead, we expect disasters linked to climate change to give rise to discourses around interdependence, as the social, economic, and ecological matrix from which independent modes of self-construal have arisen begin to erode. In the face of such challenges and changes, how long will policy in the twenty-first century continue to issue the call for "one who acts freely and with self-responsibility, self-reliantly supporting himself"? How long will classroom pedagogies emphasize showers of personalized praise and individualized learning models? How long will it still be meaningful to measure happiness and rank countries according to questions like "I am satisfied with my life" and "I have acquired the things I want in life"? How long will research continue to promote twentieth-century modes of happiness and well-being?

# References

Arnocky, S., Stroink, M., & DeCicco, T. (2007). Self-Construal Predicts Environmental Concern, Cooperation, and Conservation. *Journal of Environmental Psychology, 27*, 255–264.

Brewer, M. B., & Gardner, W. L. (1996). Who Is This "We"? Levels of Collective Identity and Self-Representations. *Journal of Personality and Social Psychology, 71*(1), 83–93.

Chuang, Y., Xie, X., & Liu, C. (2016). Interdependent Orientations Increase Pro-Environmental Preferences When Facing Self-Interest Conflicts: The Mediating Role of Self-Control. *Journal of Environmental Psychology, 46*, 96–105.

Commission for the Design of 21st Century Japan. (1999). *Report of the Commission for the Design of 21st Century Japan.* Prime Minister's Office. https://www.kantei.go.jp/jp/21century/report/pdfs/3chap1.pdf

Davis, A. C., & Stroink, M. L. (2016). Within-Culture Differences In Self-Construal, Environmental Concern, and Proenvironmental Behavior. *Ecopsychology, 8*, 64–73.

Dynes, R. (2006). Social Capital: Dealing with Community Emergencies. *Homeland Security Bulletin.* https://www.hsaj.org/resources/uploads/2022/05/2.2.5.pdf

Furuichi, K. (2011). Zentsubo no Kuni no Koufukuna Wakamono Tachi [*Happy Youth in a Hopeless Land*]. Kodansha Publishers.

Gardner, W. L., Gabriel, S., & Lee, A. Y. (1999). "I" Value Freedom, but "We" Value Relationships: Self-construal Priming Mirrors Cultural Differences in Judgment. *Psychological Science, 10*(4), 321–326.

Geertz, C. (1973). *The Interpretation of Cultures.* Basic Books.

Genda, A. (2001) Shigoto no naka no aimaina fuan—yureru wakamono no genzai [*Contemporary Youth: Swaying under the Ambiguous Instability of Employment*]. Chuokoron Shinsha (Chuokoron Publishers).

Hamamura, T. (2012). Are Cultures Becoming Individualistic? A Cross-Temporal Comparison of Individualism-Collectivism in the U.S. and Japan. *Personality and Social Psychological Review, 16*, 3–24.

Horige, I. (2013). Shinsai ga shukanteki well-being ni ateru eikyo nit suite, gendai jin no kokoro no yukue [*The Impact of Disasters on Subjective Well-Being—The State of Contemporary Minds in Japan*]. Human Interaction.

Ishino, T., Kamesaka, A., Murai, T., & Ogaki, M. (2012). Effects of the Great East Japan Earthquake on Subjective Well-Being. *Journal of Behavioral Economics and Finance, 5*, 269–272.

Jung, H. J., & Ahn, J. (2021). South Korean Education Under Psychocultural Globalization. *Ethos, 49*(1), 3–10.

Kalmus, P. (2021, November 4). Climate Depression Is Real. And It Is Spreading Fast Among Our Youth. *The Guardian.* https://www.theguardian.com/commentisfree/2021/nov/04/climate-depression-youth-crisis-world-leaders

Kasulis, T. (1998). *Intimacy or Integrity: Philosophy and Cultural Difference.* University of Hawaii Press.

Kikuike, S. (2015). Shogakko hatsu! Hitori-hitori ga kagayaku home kotoba no shawa [*Starting in Elementary School! The Shower of Praise to Make Each Individual Child Shine*]. Takobon.

Kimball, M., Levy, H., Ohtake, F., & Tsutsui, Y. (2006). *Unhappiness after Hurricane Katrina (No. w12062).* National Bureau of Economic Research.

Kitayama, S., & Markus, H. (2000). The Pursuit of Happiness and the Realization of Sympathy: Cultural Patterns of Self, Social Relations, and Well-Being. In E. Diener & E. M. Suh (Eds.), *Culture and Subjective Well-Being* (pp. 113–161). MIT Press.

Komatsu, H., & Rappleye, J. (2020). Reimagining Modern Education: Contributions from Modern Japanese Philosophy and Practice? *ECNU Review, 3*(1).

Komatsu, H., & Rappleye, J. (2021). Nihon kyoiku ha dame jyanai [Japanese Education Isn't That Bad]. Tokyo: Chikuma.

Komatsu, H., Rappleye, J., & Silova, I. (2019). Culture and the Independent Self: Obstacles to Environmental Sustainability? *Anthropocene, 26*, 100198.

Komatsu, H., Silova, I., & Rappleye, J. (2020). Will Education Post-2015 Move Us Toward Environmental Sustainability? In A. Wulff (Ed.), *The Big Book on SDG 4* (pp. 297–321). Brill.

Komatsu, H., Rappleye, J., & Silova, I. (2021). Student-Centered Learning and Sustainability: Solution or Problem? *Comparative Education Review, 65*(1), 6–33.

Komatsu, H., Rappleye, J., & Uchida, Y. (2022a). Is Happiness Possible in a Degrowth Society? *Futures, 144*, 103056 (Online First).

Komatsu, H., Rappleye, J., & Silova, I. (2022b). *Environment and Educational Sustainability: Culture Matters.* Online First. https://www.emerald.com/insight/content/doi/10.1108/JICE-04-2022-0006/full/html

Kotozaki, Y., & Kawashima, R. (2012). Effects of the Higashi-Nihon Earthquake: Posttraumatic Stress, Psychological Changes, and Cortisol Levels of Survivors. *PLoS One, 7*(4), e34612.

Kyutoku, Y., Tada, R., Umeyama, T., Harada, K., Kikuchi, S., Watanabe, E., et al. (2012). Cognitive and Psychological Reactions of the General Population Three Months after the 2011 Tohoku Earthquake and Tsunami. *PLoS One, 7*(2), e31014.

Martinsson, P., Myrseth, K. O. R., & Wollbrant, C. (2012). Reconciling Prosocial vs. Selfish Behavior: On the Role of Self-Control. *Judgment and Decision making, 7*, 304.

Metcalfe, R., Powdthavee, N., & Dolan, P. (2011). Destruction and Distress: Using a Quasi-Experiment to Show the Effects of the September 11 Attacks on Mental Well-being in the United Kingdom. *The Economic Journal, 121*, F81–F103.

Ministry of Health, Labor and Welfare. (2013). Kosei roudo hakusho [*White Paper on Labor and Welfare*]. Ministry of Health, Labor, Welfare.

Minkov, M., Dutt, P., Schachner, M., Morales, O., Sanchez, C., Jandosova, J., et al. (2017). A Revision of Hofstede's Individualism-Collectivism Dimension: A New National Index from a 56-Country Study. *Cross Cultural & Strategic Management, 24*(3), 386–404.

Nakatani, G. (2008). Shihonshugi ha naze jikaishita no ka: "nihon" saisei he no teigen [*Why Did Capitalism Unravel Itself? A Proposal for the Rebuilding of "Japan"*]. Shueisha International Publishers.

New Development Paradigm Initiative. (2014). *Towards a New Development Paradigm*. Royal Government of Bhutan. http://www.nowforourturn.org/Reframing/Towards-a-New-Development-ParadigmBhutan.pdf

Nieto, J., Carpintero, Ó., & Miguel, L. (2018). Less than 2°C? *An Economic-Environmental Evaluation of the Paris Agreement, Ecological Economics, 146*, 69–84.

Nisbett, D. (2003). *The Geography of Thought: How Asians and Westerns Think Differently … and Why*. Free Press.

Nishimoto, J., & Inoue, K. (2004). Shinzai go no shinnriteki hennka: jinseikan wo chushin toshita kentou [Psychological Changes in the Wake of Earthquakes: An Examination Centered on Changes in Worldview]. Jibun Ronkyu [*Humanities Review: The Journal of the Literary Association of Kwansei Gakuin University*], *54*, 72–86.

OECD. (2018). *OECD Environmental Outlook to 2050: The Consequences of Inaction*. OECD. https://www.oecd.org/g20/topics/energy-environment-green-growth/oecdenvironmentaloutlookto2050theconsequencesofinaction.htm

Ogihara, Y., & Uchida, Y. (2014). Does Individualism Bring Happiness? Negative Effects of Individualism on Interpersonal Relationships and Happiness. *Frontiers in Psychology, 5*, 135.

Rappleye, J., & Komatsu, H. (2020). Towards (Comparative) Educational Research for a Finite Future. *Comparative Education, 56*(2), 190–217.

Seeley, E. A., & Gardner, W. L. (2003). The "Selfless" and Self-Regulation: The Role of Chronic Other-Orientation in Averting Self-Regulatory Depletion. *Self and Identity, 2*, 103e117.

Toivonen, T., Norasakkunkit, V., & Uchida, Y. (2011). Unable to Conform, Unwilling to Rebel? Youth, Culture, and Motivation in Globalizing Japan. *Frontiers in Cultural Psychology, 2*, 207.

Trafimow, D., Triandis, H. C., & Goto, S. G. (1991). Some Tests of the Distinction between the Private Self and the Collective Self. *Journal of Personality and Social Psychology, 60*(5), 649–655.

Uchida, Y., & Norasakkunkit, Y. (2015). The Neet and Hikikomori Spectrum: Assessing Risks and Consequences of Becoming Culturally Marginalized. *Frontiers of Psychology, 6*, 1117.

Uchida, Y., Takahashi, Y., & Kawahara, K. (2014). Changes in Hedonic and Eudaimonic Well-Being after a Severe Nationwide Disaster: The Case of the Great East Japan Earthquake. *Journal of Happiness Studies, 15*(1), 207–221.

United Nations. (2021). *Nationally Determined Contributions Under the Paris Agreement.* United Nations Climate Change.

Yamada, S. (2009). Shinbyoudo shakai: kibo kakusa wo koete [*New Social Equality: Overcoming Aspirational Inequalities*]. Bungeishunjyu.

Zielenziger, M. (2007). *Shutting out the Sun: How Japan Created its Own Lost Generation.* Vintage.

# 7

# Conclusion

Our aim in this volume has been to raise awareness around an alternative approach to happiness and well-being. Against the backdrop of a departure from the twentieth century's GDP = Happiness paradigm, we sought to lay out the basic concepts, outline the general policy storyline, and present supporting evidence in support of this alternative. We tried to keep the presentation concise, and accessible to a wide audience, in hopes of attracting many readers who might not otherwise encounter such ideas. Inevitably, brevity leads to many places wanting of deeper elaboration. Readers seeking more details and nuance may wish to now turn to our many scholarly articles published primarily in the fields of (cultural) psychology, sociology, and education, but increasingly in environmental studies as well. Given the nature of scholarly publishing, the larger story that we have outlined herein tends to get lost among the many smaller empirical studies that support it. On the other hand, outlines of the larger plot can tend to sound stereotypical, essentialist, homogenizing, or quaint, as when we gesture toward, say, 'harmony'. We hope to have persuaded readers we are not engaged in this sort of shallow work. Instead, we are trying to contribute to a discussion around alternatives. We are interested neither in relativism nor in competition, but instead dedicated to collective search for new solutions to shared problems. Our stance is consistent with an interdependent mode.

© The Author(s) 2024
Y. Uchida, J. Rappleye, *An Interdependent Approach to Happiness and Well-Being*,
https://doi.org/10.1007/978-3-031-26260-9_7

For readers persuaded by our account, it is useful to conclude the volume by thinking about the future of this line of work. How to continue developing and nuancing the core conceptions, while also working to expand their explanatory power across space and academic disciplines? How to raise awareness around it, and bring it further into mainstream policy and practice discourses, without losing necessary nuance?

An initial, and yet crucial step, is to move beyond Japan. As we have underscored, the interdependent pattern is foregrounded strongly across Japan. Yet, it is not unique to Japan. We surmise that much of East Asia shares these general patterns, both in self-construal and cultural contexts that reinforce it. Returning to some of the data we presented in Chap. 5 (Figs. 5.9, 5.10), across East Asia we notice a strikingly similar pattern to Japan: well-being scores on an individualized measure like Cantril's Ladder are low across East Asia, but when asked "Are you happy at school?", the same East Asian students score above the average. We hypothesize that at work behind such a paradox is an interdependent orientation shared across much of East and Southeast Asia. Despite very different political, social, economic, developmental, and religious contexts among these countries, there appears to be a shared orientation. Japan, in avoiding Western colonization and being relatively 'late' in adopting cultural systems favoring independence (e.g., student-centered learning, pedagogies for self-esteem), may show more pronounced *but not necessarily unique* results. We surmise that much of East and Southeast Asia continues to live largely in an interdependent mode, although a second floor of subjectivity continues to be slowly erected, constructed by Western institutions, discourses, and pedagogies. Future research should move beyond Japan to engage these different contexts and layers, to test the wider viability of the concept and the interplay between the independent and interdependent modes across these diverse societies. This work will teach us much more about interdependence.

Building on this, the next step is to move beyond psychology, linking up with the social sciences and pursuing a trans-disciplinary approach. As discussed in Chap. 5, independent modes of self-construal and happiness are only held in place by meanings, practices, products, and institutions existing in the social world. Part of the shift that cultural psychologists have demanded of mainstream psychology is a shift in focus away from cognition and contents of 'culture', toward the process, mechanisms, maintenance, and modification of those patterns within a cycle of co-construction (Markus & Kitayama, 2010; Adams & Markus, 2004). In the not-so-distant past, sociology had a strong focus on culture, the ways it linked to cognition, and a sense of diversity in these patterns (e.g., Berger & Luckmann, 1966; Eisenstadt, 1996). Unfortunately, under the sway of universalist narratives of historical materialism and poststructuralism, many sociologists have turned away from culture in the sense we have been describing it here. Recall that in Chap. 4 we briefly rehearsed a similar turn away from 'culture' in anthropology. When sociologists do discuss culture, they often give the impression that difference has been swept away by the last few decades of globalization (e.g., Meyer et al., 1997). A more fruitful direction is, in our opinion, to reconnect psychology and the social sciences in a joint project around explicating patterns of culture. Recently, a world-leading sociologist and psychologist jointly raised a similar call:

> Psychology and sociology typically have different endgames and thus social psychologists of different disciplinary persuasions sail past each other in the night. Yet the current notable convergence between psychological and sociological social psychologists in definitions and approaches to culture, as well as their shared view that cultures and selves/identities constitute each other in a cycle of mutual constitution, suggests that the time may be right for sustained interdisciplinary work. Psychologists could benefit from sociological theorizing on roles, networks, institutions, and on how ideas and practices diffuse and cultures change. Sociologists could benefit from psychological research on when and how specific psychological tendencies vary with specific features of context. (DiMaggio & Markus, 2010)

In the current volume, we have tried to answer this call. We sought to show how, although often viewed as incommensurable from a traditional disciplinary perspective, the Protestant Ethic of Weber, the Zen-inspired philosophies of Nishida Kitaro, traditional Japanese educational practices, new pedagogies of self-esteem, and even the World Happiness rankings all comprise a surprisingly coherent overall story. Through engaging in such trans-disciplinary thinking, we not only breathe new life into existing disciplinary debates, but recover some of the ground lost by the overspecialization of scholarship.

Building on these steps—wider regional findings and trans-disciplinarity—the next move needs to be greater engagement with global organizations. Despite their global reach and rhetoric of inclusion, the leading international organizations—the OECD, World Bank, UNICEF, and UNESCO—remain decidedly narrow in their outlook. This has been a long-standing problem for the United Nations, but is perhaps inevitable in the world of policymaking that continues to take its cues solely from Western scholarship. In the past, when these global organizations were focused primarily on non-cultural issues such as, say, the measuring of GDP growth rates, building bridges, getting children into schools, ensuring their basic health, and comparing math achievement scores, the problems were less apparent. Yet, the rise of psycho-cultural globalization witnessed in, say, measuring the happiness of a population (WHR), measuring student well-being (PISA, UNICEF), and promoting Happy Schools (UNESCO) brings into stark relief the tension between cultural differences in culture and drive for universal solutions still found among these organizations.

We note some welcome movement in, say, the inclusion of the first Global Survey on Harmony and Balance included in the WHR 2022, and invitations for us to contribute to the work of UNESCO and UNICEF (see Rappleye & UNESCO, 2022; Uchida & Rappleye, 2022). But there needs to be a stronger concerted push to gain recognition of these alternatives in global organizations. To the degree to which an interdependent approach comes to be recognized within macro global cultural products (e.g., WHR rankings, OECD reports), it will serve to reinforce the micro cultural products of national happiness indicators, pedagogical practices, and everyday conversation, lifestyles, and interactions.

If we could affect such a shift, the potential sources of intra-cultural learning would quickly diversify. In the twentieth century, cultural learning was dominated by 'learning from the West'. The shift to interdependence would shift the focus, giving rise to new reference societies. Instead of Finland, we might look to, say, Bhutan. As we have seen, Bhutan had—for a time—become a reference point in discussions of a new development paradigm. So it is worth pausing to understand the reasons for this. We ourselves have visited Bhutan and surrounding areas of the Himalayas on several occasions, and had opportunities to speak with officials there about this question. From such discussions, it is clear that the worldview and notion of happiness in Bhutan is very different from the so-called acquisition-oriented happiness of WEIRD countries. Of course, in Bhutan, getting a good education, having self-esteem, and having a strong work ethic are still recognized. However, there is an equally, if not stronger emphasis on gratitude and contentment. The people of Bhutan place great importance on feeling a sense of fulfillment in their daily lives while being grateful to their parents, ancestors, nature, animals, and the land. The Bhutanese tend to value helping others, rather than getting ahead as an individual. Bhutan is, of course, not the idyllic Shangri-la that people imagine. Economically, Bhutan is one of the poorest countries in the world, and the infrastructure in some rural areas is not well developed. However, in visiting the country one senses that people are living with pride and trust in others. When visiting Buddhist temples, we see men and women of all ages sitting for long hours praying. When visiting schools, we see children quite happily and diligently studying. The forest is deep in color. The flow of time was slow and unhurried (the Bhutanese way). Perhaps because the concept of 'reincarnation' is deeply rooted in Bhutan, time seems to be understood differently there as well: even if we do not succeed in this life, we may do well in the next. There is a sense that one's happiness is connected not only to the person or task directly in front of you, but also to your ancestors and descendants that you may never meet in the future. It is a perspective that helps us—all of us—think about happiness in a less hurried, one-off manner, and over a long period of time. Indeed, the whole complex of Bhutanese Buddhism underpins much of their approach to GNH.

The policy of Gross National Happiness (GNH) is purported to have originated in the 1970s when the fourth king of Bhutan said, "I want Bhutan to be a country that values GNH more than GDP". The GNH policy turned out to be a great opportunity for the small country of Bhutan to make a worldwide contribution. Bhutan has helped the world see that if economic growth does not make people happy, then perhaps slow growth is one option; perhaps it is better to value what inevitably comes to be lost through economic growth. It is indeed quite revolutionary to formulate policies that consider happiness, and to use indicators to measure and analyze them. Policies proposed by various ministries are evaluated from the perspective of GNH. If the plan does not meet this perspective, it may be rejected, which will affect the subsequent budget allocation of line ministries. This, at least to our knowledge, is unprecedented globally.

In their GNH survey, the Royal Bhutan Institute, a government think-tank, conducts subjective measurements of economic, social, environmental, and other factors. We have witnessed this first hand. The survey is large and thick, and because some people cannot read, the researchers walk through the mountains to visit the survey targets, explain the survey forms orally, and request that respondents take their time to answer all the questions. Often the exercise extends over the course of an entire day. The results of the survey are quantified to evaluate, for example, whether people are really satisfied with the forest environment, the working environment, politics, and the economy. It includes the government's four official policy pillars of 'conservation of the natural environment', 'equitable and sustainable socioeconomic development', 'good governance', and 'protection and promotion of traditional culture'. The GNH indicator itself has nine domains: 'how to consume time', 'physical health', 'psychological health and well-being', 'community activities', 'traditional culture', 'good governance', 'living standards', 'environment', and 'education'. Happiness is defined as 'met' when at least six of these criteria are met to a satisfactory level. Clearly, these criteria reflect the values of the Bhutanese context.

There is, inevitably, criticism that the GNH policy is a message from the government that people should accept their less-than-wealthy economic status and simply accept their current situation as 'happy'. However, what is actually happening on the ground in Bhutan is not an intervention that says, "Think of this as happiness", but rather concrete decision-making on what the country should do to ensure the happiness of each person, and the implementation of surveys that make this possible. To be sure, changes under globalization are also occurring in Bhutan. Thimphu, the capital, is rapidly urbanizing, and more and more young people from rural areas are leaving farming and coming to the capital. Naturally, the capital does not have the capacity to hold all the migrant workers, leading to problems like youth unemployment. With the rise of smartphones and the internet there have been changes in values and awareness. Some youth now dream of a 'happier' future in Western countries, a tension well captured in the insightful film *Lunana: A Yak in the Classroom* (2019). Under these circumstances, the world is watching to see how Bhutan's happiness will change. But whatever happens, we must recognize that Bhutan, much more than a powerful economic country like Japan, has impacted the global discourse on happiness: as pointed out in Chap. 1: the 2011 UN resolution on Happiness featured discussions of Bhutan. If such a small country can impact the discussion, we imagine that a collective global effort confirmed by rigorous empirical science can further the dialogue even that much more.

On that note, we close with a final gesture toward Japan. Specifically, we would like to raise a potential analogy between the interdependent mode and *umami*. This might be startling to some readers, but we have used a similar strategy in our other writings to leave readers with a sense of how things can and do change (Rappleye, 2020). As some readers may recall from their high school biology and life science lessons, *umami* is now recognized as one of the five basic tastes. But what you might not know is how this happened: as late as the 1980s, the scientific community was convinced there were only four basic tastes—sweet, sour, salty, and bitter. Yet a different taste—*umami*—is central to the patterns of Japanese cuisine: it is the key ingredient in everything from Japanese soup bases using seaweed to shiitake mushrooms to miso to soy sauce to high-end green teas (the higher the *umami* content, the higher the price). Some trace the development of *umami* in Japanese cuisine to the Buddhist ban on eating meat: ways to enhance flavor could not rely on meat-based sauces as found in many Western countries. Yet, *umami* is, in fact, also present in many familiar Western foods like aged beef, ripe tomatoes, and Parmesan cheese, delivering a rich, satisfying 'something' that is not captured by the other four flavors. That is, it is not unique to Japan but the wider religio-philosophical culture may have brought it to greater gastrocultural prominence there.

Nonetheless, for over a century Western researchers (and thus the global scientific community) refused to recognize *umami*. These WEIRD scientists tended to view human taste like they viewed human visual perception: basic colors derived from the universality of the nature of humans' 'basic' biologically determined perception abilities. Stubborn Japanese scientists persisted in arguing that *umami* was a distinct taste, and were reminded everyday—through experiences of drinking tea and miso soup—that it did, in fact, exist. It was only beginning in the late 1950s, when Japanese scientists began elaborating *umami* in scientific terms and publishing in English, did people start to pay attention. Still it took another 30 years—1990—until it was officially recognized as the fifth basic taste. Today, *umami* is celebrated from Paris to Los Angeles, and recognized as an element of healthier diets: it allows for flavor enhancement with less salt and fat (e.g., Mouritsen, 2012; Sasano et al., 2015). Are we really to believe that WEIRD categories capture the full range of human experience (taste)?

Through this brief closing gesture to *umami*, we seek to underscore just how little we still know, even today, about the wider world and about our-*selves*. We wonder aloud whether, one day, an interdependent approach to well-being can become recognized as another 'basic' dimension of the experience of *being* human: a sense of shared happiness rooted in balance and harmony, leading to a healthier, more sustainable twenty-first century.

# References

Adams, G., & Markus, H. R. (2004). Toward a Conception of Culture Suitable for a Social Psychology of Culture. In M. Schaller & C. S. Crandall (Eds.), *The Psychological Foundations of Culture* (pp. 335–360). Erlbaum.

Berger, P. L., & Luckmann, T. (1966). *The Social Construction of Reality: A Treatise in the Sociology of Knowledge*. Doubleday.

DiMaggio, P., & Markus, H. (2010). Culture and Social Psychology: Converging Perspectives. *Social Psychology Quarterly, 73*(4), 313–357.

Eisenstadt, S. N. (1996). *Japanese Civilization: A Comparative View*. University of Chicago Press.

Markus, H. R., & Kitayama, S. (2010). Cultures and Selves: A Cycle of Mutual Constitution. *Perspectives on Psychological Science, 5*, 420–430.

Meyer, J., Boli, J., Thomas, G., & Ramirez, F. (1997). World Society and the Nation-State. *The American Journal of Sociology, 103*(1), 144–181.

Mouritsen, O. (2012). Umami Flavour as a Means of Regulating Food Intake and Improving Nutrition and Health. *Nutrition and Health, 21*(1).

Rappleye, J. (2020). Comparative Education as Cultural Critique. *Comparative Education, 56*(1), 39–56.

Rappleye, J., & UNESCO. (2022). Happy Schools Guide and Toolkit: A Resource for Happiness, Learners' Well-Being and Social and Emotional Learning. UNESCO. https://bangkok.unesco.org/content/happy-schools-guide-and-toolkit-resource-happiness-learners-well-being-and-social-and

Rappleye, J., & Komatsu, H. (2022). Learning to Be, Differently? UNESCO, The Faure Report, and Mutual Appreciation in Retrospect and Prospect. *Knowledge Cultures, 10*(2). https://www.addletonacademicpublishers.com/contents-kc/2492-volume-10-2-2022/4266-learning-to-be-differently-unesco-the-faure-report-and-mutual-appreciation-in-retrospect-and-prospect

Sasano, T., Satoh-Kuriwada, S., & Shoji, N. (2015). The Important Role of Umami Taste in Oral and Overall Health. *Flavour, 4*, 10.

Uchida, Y., & Rappleye, J., (2022, September 1). High-Level Discussion with UNICEF: Subjective Well-Being and Future Collaboration (Online, with Amanda Marlin, Camila Texeira, Gwyther Rees).

# Index[1]

---

[1] Note: Page numbers followed by 'n' refer to notes.

Y. Uchida, J. Rappleye, *An Interdependent Approach to Happiness and Well-Being*,
https://doi.org/10.1007/978-3-031-26260-9

# GPSR Compliance

*The European Union's (EU) General Product Safety Regulation (GPSR) is a set of rules that requires consumer products to be safe and our obligations to ensure this.*

*If you have any concerns about our products, you can contact us on ProductSafety@springernature.com*

In case Publisher is established outside the EU, the EU authorized representative is:

Springer Nature Customer Service Center GmbH
Europaplatz 3
69115 Heidelberg, Germany

Zeitfracht Medien GmbH
Ferdinand-Jühlke-Straße 7
99095 Erfurt, Deutschland
produktsicherheit@kolibri360.de